Supplementary Cementing Materials in Concrete

Supplementary Cementing Materials in Concrete

Michael Thomas

CRC Press
Taylor & Francis Group
Boca Raton   London   New York

CRC Press is an imprint of the
Taylor & Francis Group, an **informa** business

CRC Press
Taylor & Francis Group
6000 Broken Sound Parkway NW, Suite 300
Boca Raton, FL 33487-2742

First issued in paperback 2017

Version Date: 20121128

ISBN 13: 978-1-138-07565-8 (pbk)
ISBN 13: 978-1-4665-7298-0 (hbk)

Library of Congress Cataloging-in-Publication Data

Thomas, Michael D. A.
    Supplementary cementing materials in concrete / Michael Thomas.
    pages cm
    Includes bibliographical references and index.
    ISBN 978-1-4665-7298-0 (hardback)
    1. Concrete--Additives. I. Title.

TP884.A3T47 2013
666'.893--dc23   2012035470

Visit the Taylor & Francis Web site at
http://www.taylorandfrancis.com

and the CRC Press Web site at
http://www.crcpress.com

Contents

Preface

Supplementary cementing materials (SCMs) are widely used in the production of concrete worldwide. The term covers a broad range of materials, including natural (and manufactured) pozzolans and industrial by-products, such as fly ash, slag, and silica fume. Although the properties of the different types of SCM vary considerably, they share the ability to react chemically in concrete and produce cementitious compounds that *supplement* those produced by the hydration of portland cement. Appropriately used, SCMs can improve many of the fresh and hardened properties of concrete. The use of SCMs in concrete is not new, as the use of pozzolans, such as volcanic ash, has been traced back to ancient Greek and Roman civilizations, and many of the structures built using pozzolans are still standing today.

This book provides detailed information on the use of supplementary cementing materials in concrete, including the provenance of these materials and their chemical, physical, and mineralogical properties. By providing an understanding of the chemical reactions involving pozzolans and slag, and the resulting changes in the microstructure of the concrete, it is explained how SCMs influence the mechanical properties of concrete and improve its durability. In the book, it is explained how various SCMs influence hydration reactions and the evolution of the pore structure and pore solution composition. Results are presented from a wide range of studies to demonstrate the impact of SCMs on the temperature rise in concrete, the development of strength, and volume stability. The author uses his own experience to demonstrate that the proper use of SCMs can significantly reduce the permeability of concrete and substantially increase its resistance to the penetration of deleterious agents such as chlorides, thereby extending the service life of concrete in aggressive environments. Data are also presented to demonstrate the effect of SCM on a wide range of deterioration processes, including corrosion of embedded steel reinforcement, carbonation, alkali-silica reaction, freeze-thaw damage and de-icer salt scaling, sulfate attack, and delayed ettringite formation.

However, SCMs are not a panacea for concrete, and improper use may be injurious to certain properties. Achieving the maximum benefit from

SCMs requires an understanding of the materials and how they impact concrete properties under various conditions. It is hoped that this book will help engineers and practitioners optimize the use of supplementary cementing materials to improve concrete performance.

About the Author

Michael Thomas is a professor in the Department of Civil Engineering at the University of New Brunswick (UNB) and a registered professional engineer in the province of New Brunswick. He has been working in the field of cement and concrete research since 1983. Prior to joining UNB in 2002 he had been on the faculty at the University of Toronto since 1994, and previous to this he worked as a concrete materials engineer with Ontario Hydro in Canada and as a research fellow with the Building Research Establishment in the United Kingdom.

Dr. Thomas's main research interests are concrete durability and the use of industrial by-products, including pozzolans and slag. His studies on durability have included alkali-silica reaction, delayed ettringite formation, sulfate attack, de-icer salt scaling, carbonation, chloride ingress, and embedded steel corrosion. He is also active in the area of service life modeling, and in the repair and maintenance of concrete structures. He has authored more than 200 technical papers and reports on these subjects, and is a coauthor of the service life model, Life-365.

Dr. Thomas is active on technical committees within the American Concrete Institute (ACI), American Society of Testing and Materials (ASTM), Reunion Internationale des Laboratoires et Experts des Materiaux (RILEM), and the Canadian Standards Association. He was a recipient of the ACI's Wason Medal for Materials Research in 1997 and 2009, the ACI Construction Practice Award in 2001, and was elected to a fellow of the institute in 2006. He is also a fellow of the Institute of Concrete Technology.

Chapter 1

Introduction

The appropriate use of suitable supplementary cementing materials (SCMs) can lead to many improvements in the fresh and hardened properties of concrete, including reduced water demand, better workability, increased long-term strength, and improved durability in aggressive environments. However, if SCMs are not used properly or poor quality SCMs are used, the fresh and hardened concrete properties may be adversely affected. This textbook provides detailed descriptions of the different SCMs available, the chemical reactions that they undergo in concrete, the impact of these reactions on the development of the microstructure in concrete, and how the use of SCMs influences the properties of concrete.

1.1 WHAT ARE SCMS?

The term *supplementary cementing materials* (SCMs) defines a broad range of materials that are widely used in concrete in addition to portland cement.[1] An SCM may be defined as "a material that, when used in conjunction with portland cement, contributes to the properties of the hardened concrete through hydraulic or pozzolanic activity, or both" (CSA A3001, 2003). As such SCMs include both pozzolans and hydraulic materials. SCMs that are commonly used in concrete today include fly ash, ground granulated blast furnace slag, silica fume, and a wide range of natural pozzolans, such as volcanic ash, calcined clay or shale, and diatomaceous earth.

A pozzolan is defined as "a siliceous or siliceous and aluminous material that in itself possesses little or no cementitious value but that will, in finely divided form and in the presence of moisture, chemically react with calcium hydroxide (a hydration product of portland cement) at ordinary temperatures to form compounds having cementitious properties" (ACI

---

[1] *Supplementary cementing materials* is now the preferred term for the class of materials that used to be called mineral admixtures.

116, 2000). Pozzolans that are commonly used in concrete include fly ash, silica fume, and a variety of natural pozzolans, such as calcined clay and shale, and volcanic ash. Some SCMs, such as ground granulated blast furnace slag, show hydraulic behavior in that they react with water directly to form hydration products with cementitious properties. Such materials do not require portland cement to harden and gain strength, but the chemical reactions are greatly accelerated by the presence of portland cement, and thus materials such as slag are most often used in combination with portland cement. Some materials, such as fly ashes that contain a high amount of calcium (Class C fly ashes) will react both pozzolanically and hydraulically when used in concrete.

SCMs undergo chemical reactions in concrete, and the products of reaction are cementitious in nature; that is, the products help bind the components of the concrete together in the same manner as the reaction (or hydration) products of portland cement. Consequently, SCMs are considered to be part of the cementing material component of the concrete, and they should be included in the calculation of the water-to-cementing materials ratio, W/CM, of the concrete. As such, SCMs should be differentiated from finely divided mineral fillers, such as ground limestone or quartz, which are generally chemically inert in concrete and are not considered to be part of the cementing material.

SCMs may be added as a separate ingredient to the concrete mixer or as a component of a blended cement; a blended cement is a manufactured product consisting of portland cement blended or interground with one or more SCMs.

1.2 HISTORY OF USE

It is well known that the use of natural pozzolans, such as volcanic ash, to produce concrete dates back more than two millennia to the ancient civilizations of Greece and Rome, although it has been claimed that the oldest example of a hydraulic binder dates back more than six millennia and consisted of a mixture of lime and a diatomaceous earth from the Persian Gulf (Malinowski and Frifelt, 1993). The Greeks produced hydraulic binders from mixtures of lime and volcanic ash from the Island of Santorini (Thera), and archaeologists have uncovered concrete structures built as far back as 600 B.C. using these pozzolan-lime mixes (Efstathiadis, 1978). The Greeks passed the technology on to the Romans in about 150 B.C., and the Romans developed a wide range of pozzolans throughout their empire during their 600-year domination (ACI 232, 2012), including Rhenish trass that was used throughout Germany and a source of volcanic ash from the village of Pozzouli, near Italy, which lent its name to the materials we now call pozzolans. In 13 B.C., the Roman engineer Vitruvius, in his handbook for architects, described how to use pozzolan and lime to obtain materials "that hardened in air and underwater" (BCA, 1999) and "that neither waves could break, nor water dissolve" (ACI 232, 2012). Many of the structures built using pozzolan-lime binders are still standing today, and perhaps the most impressive of these is the Pantheon in Rome. The Pantheon is not only still standing, but it is still used today for the same purpose for which it was built. The Pantheon was built under the direction of Emperor Hadrian around 120 A.D. The walls of the Pantheon are 6 m (20 ft) thick and are built with concrete consisting of tuff and pozzolana, but it is the 43-m (141-ft) -diameter dome that is most remarkable, the top portion being composed of lightweight concrete with pumice aggregate bound together by a pozzolan-lime binder (Idorn, 1997).

Since the advent of portland cement (which was first patented in 1824), pozzolan-lime binders have been replaced by mixtures of portland cement and natural pozzolans, which have been used in structures around the world during the last century or so. The first major use in North America of a natural pozzolan in portland cement concrete was in the construction of the Los Angeles aqueduct in 1910–1912 (Price, 1975), and natural pozzolans have since been used in the construction of a number of large dams in the United States.

Although the first experiments on slag-lime mortars were performed by Loirot in 1774 (Mather, 1957), the first commercial use of slag-lime cements was in Germany in 1865 (ACI 233, 2003). One of the first major uses of slag-lime cements was in the construction of the Paris metro beginning in 1889 (Thomas, 1979). Blended slag-portland cements were first used in Germany in 1892 and later in the United States in 1896 (ACI 233, 2003); slag-portland cement mortar was used in the masonry mortar for the construction of the Empire State Building, which started in 1930 (Prusinski, 2006). It has been estimated that by 1981 nearly 20% of the total hydraulic cement produced in Europe was blended cement containing slag and portland cement. Since gaining acceptance in the late 1950s, the use of slag as a separate cementing material added at the mixer to portland cement concrete has grown and the majority of slag is used in this manner in North America today. The U.S. Slag Cement Association reported that the total amount of slag shipped (by the association's members) in the United States in 2006 was just over 3.6 Mt (4.0 million U.S. tons), with 3.3 Mt (3.6 million tons) of this being shipped for use as a separated product and the remainder being shipped as a blended cement (SCA, 2006).

The potential for using fly ash as a supplementary cementing material in concrete has been known almost since the start of the last century (Anon., 1914); however, fly ash from coal-burning electricity-generating plants did not become widely available until the 1930s (ACI 232, 2003) and it wasn't until the mid-1900s that significant utilization of fly ash in concrete began, following the pioneering research conducted at the University of

California, Berkeley (Davis et al., 1937). The first major use of fly ash in the United States was in the construction of the Hungry Horse Dam in Montana, which was built between 1948 and 1952 (USBR, 1948). The last 50 years has seen the use of fly ash in concrete grow dramatically, with close to 13.6 Mt (15 million U.S. tons) used in concrete, concrete products, and grouts in the United States in 2005 (ACAA, 2006). Fly ash is by far the most widely used SCM in North America, with a recent estimate showing that just over half the concrete produced in the United States contains some level of fly ash (Thomas, 2007). In North America, the vast majority of the fly ash used in concrete is added at the mixer as a separate ingredient, and only a very small proportion of the fly ash used is added as a component of blended cement. The use of blended cements containing fly ash (or other SCMs) is more common in Europe.

Silica fume is a relatively recent SCM. Its potential use as a concrete admixture was discovered by a Norwegian researcher in 1948 (Bernhardt, 1952), and although its first field application came shortly thereafter in 1952 (Fiskaa, 1973), its first major use in structural concrete in Scandinavia did not occur until 1971 (Radjy et al., 1986). One of its first appearances in the North American market was as a component in a bagged product for industrial applications where concrete was exposed to aggressive compounds such as ammonium nitrates (Thomas et al., 1998). Its first commercial use in the United States was in 1983, where silica fume concrete was used for its high abrasion resistance for the repair of a stilling basis at the Kinzua Dam in Pennsylvania (Holland et al., 1986). In Canada, the use of silica fume in ready-mix concrete began in 1983 (Skrastins and Zoldners, 1983), and the first blended silica fume cement was produced in 1982 (Thomas et al., 1998). In 1986 silica fume was used together with slag to produce high-strength concrete for the 68-story Scotia Plaza in Toronto (Ryell and Bickley, 1987). Silica fume is now widely used in high-strength and high-performance concrete throughout North America. In the United States the trend has been to add silica fume as a separate ingredient to the concrete mixer; however, in Canada silica fume is usually added to a concrete as a blended cement (Thomas et al., 1998).

Chapter 2

Origin and nature of SCMs

2.1 GENERAL

The chemical and physical properties of supplementary cementing materials (SCMs), the chemical reactions that occur with these materials, and the impact that they have on concrete vary widely among the different types of SCM. Table 2.1 describes the nature of five commonly used SCMs: low-calcium fly ash, high-calcium fly ash, slag, silica fume, and metakaolin (thermally-activated kaolin clay). The chemical composition of particular samples of these same SCMs is shown in Table 2.2. Figure 2.1 shows electron micrographs of fly ash, slag, silica fume, and metakaolin. The chemical and mineralogical composition, the particle size, and the morphology of the particles vary significantly among SCMs. Both silica fume and fly ash are predominantly comprised of spherical particles, but the average particle size of silica fume is about 100 times finer than that of fly ash. Slag and metakaolin both require grinding to make the material suitable for use in concrete, and this produces angular particles. The fineness of these materials depends on the extent of grinding. Table 2.2 shows typical chemical compositions for different SCMs, and the wide variation between the materials is shown on a CaO-SiO_2-Al_2O_3 ternary plot in Figure 2.2. Silica fume is almost entirely composed of silica (typically >90% SiO_2) and is a purely pozzolanic material with no hydraulic properties. Slag, on the other hand, has a composition relatively close to that of portland cement (but is glassy rather than crystalline) and is hydraulic in nature with little or no pozzolanic behavior. Table 2.3 shows how, generally, the hydraulicity of an SCM increases with its calcium content. Table 2.1 and Figure 2.2 show how the composition of fly ash can vary considerably, ranging from materials with very low calcium contents (<1% CaO) that only react pozzolanically to materials with much higher calcium contents (>30% CaO) that display both pozzolanic and hydraulic behavior.

The considerable variation in properties between the different types of SCM, and even within a single type (such as fly ash), means that the impact

Table 2.1 Nature of some SCMs

Property	Low-CaO fly ash	High-CaO fly ash	Slag	Silica fume	Metakaolin
Chemical composition (see Table 2.2)					
Mineralogy	Al-Si glass, inert crystalline phases	Ca-Al-Si glass, some Ca-Al glass, crystalline C_2S, C_3S, $C\hat{S}$, MgO, and lime	Ca-Al-Si-Mg glass	Silicate glass	Al-Si in non-crystalline form
Shape	Spherical	Spherical	Angular crushed particles	Spherical	Angular porous particles
Median size (μm)	5–20	2–20	5–20	0.1–0.2	1–2
Surface area (m²/kg)	300–500	300–500	400–650	15,000–25,000	
S.G.	1.9–2.8	1.9–2.8	2.85–2.95	2.2–2.3	2.5
Bulk density (kg/m³)		1200		130–430 (undensified)	
Nature of reaction	Pozzolanic	Pozzolanic and hydraulic	Hydraulic	Pozzolanic	Pozzolanic
Color	Gray	Gray to buff white	White	Dark gray to black	White

Source: Data from Mehta, P.K., in *Proceedings of the First International Conference on the Use of Fly Ash, Silica Fume, Slag and Other Mineral By-Products in Concrete*, ACI SP-79, Vol. I, America Concrete Institute, Detroit, 1983, pp. 1–46.

Table 2.2 Chemical composition of some typical SCMs

	Low-calcium fly ash	High-CaO fly ash	Slag	Silica fume	Metakaolin
SiO_2	56	32	36	97	52
Al_2O_3	28	18	10	0.52	45
Fe_2O_3	6.8	5.2	0.5	0.14	0.6
CaO	1.5	30	35	0.58	0.05
SO_3	0.1	2.6	3.5	0.01	0.00
MgO	0.9	5.2	14	0.13	0.00
Na_2O	0.4	1.2	0.35	0.04	0.21
K_2O	2.4	0.2	0.48	0.42	0.16
LOI	2.8	0.6	+1.7[a]	1.5	0.51

[a] Slags increase in mass during ignition due to the oxidation of sulfides present in the slag.

Figure 2.1 Scanning electron microscope images of fly ash (top left), slag (top right), silica fume (bottom left), and calcined clay (bottom right). (From Kosmatka, S.H., and Wilson, M.L., *Design and Control of Concrete Mixtures*, EB001, 15th ed., Portland Cement Association, Skokie, IL, 2011. Printed with permission from the Portland Cement Association.)

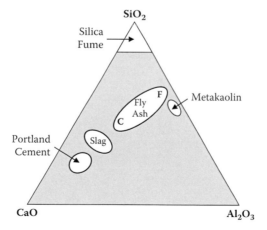

Figure 2.2 Chemical composition of some commonly used SCMs.

Table 2.3 Pozzolanic and hydraulic reactions of SCMs

SCM	Pozzolanicity	Hydraulicity	Calcium content
Silica fume	XXXXX		Low (<1%)
Metakaolin	XXXXX		
Low-CaO fly ash	XXXX		$\downarrow$
High-CaO fly ash	XXX	XX	
Slag	X	XXXX	High (>30%)

of SCMs on concrete properties varies widely, and that there are different requirements for their use in concrete.

2.2 FLY ASH

2.2.1 Production of fly ash

Fly ash is a by-product of burning coal in an electrical generating station. Figure 2.3 shows a typical layout of a coal-burning generating station. Coal is first pulverized in grinding mills before being blown with air into the burning zone of the furnace. In this zone the combustible constituents of coal (principally carbon, hydrogen, and oxygen) ignite, producing heat with temperatures reaching approximately 1500°C (2700°F). At this temperature the noncombustible inorganic minerals associated with the coal (such as quartz, calcite, gypsum, pyrite, feldspar, and clay minerals) melt and form small liquid droplets. These droplets are carried from the burning zone with the flue gases and cool rapidly, forming small spherical

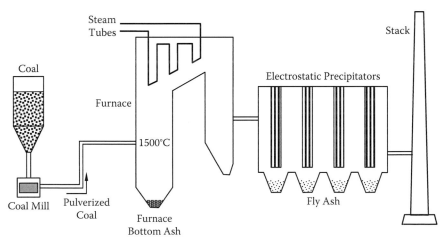

Figure 2.3 Schematic layout of a coal-fired electrical generating station.

glassy particles as they leave the chamber. These solid particles are collected from the flue gases using mechanical and electrical precipitators, or baghouses. The particles of ash that "fly" away from the furnace with the flue gases are called fly ash, and this material can be used as an SCM in portland cement concrete. Fly ash is also called pulverized fuel ash (PFA) in some countries. Heavier unburned ash particles drop to the bottom of the furnace, and this material is termed bottom ash or furnace bottom ash; such material is not generally suitable for use as a cementitious material for concrete, but is used in the manufacture of masonry block.

2.2.2 Composition and properties of fly ash

The performance of fly ash in concrete is strongly influenced by its physical, mineralogical, and chemical properties. The mineralogical and chemical compositions are dependent to a large extent on the composition of the coal, and since a wide range of domestic and imported coals (anthracite, bituminous, subbituminous, and lignite) are burned in different generating stations in North America, the properties of the fly ash can be very different between sources and collection methods. The burning conditions within a power plant can also affect the properties of the fly ash.

Table 2.1 shows the chemical composition of two different fly ashes from sources within the United States. The low-calcium fly ash results from the burning of an eastern bituminous coal, and the high-calcium fly ash was produced from a subbituminous coal from the Powder River Basin. The range of fly ash composition encountered in North America is shown in Figure 2.2, with calcium contents ranging anywhere from less than 1% CaO to more than 30% CaO.

The calcium content of the fly ash is perhaps the best indicator of how the fly ash will behave in concrete (Thomas et al., 1999), although other compounds, such as the alkalis (Na_2O and K_2O), carbon (usually measured as LOI), and sulfate (SO_3), can affect the performance of the fly ash. Low-calcium fly ashes (<8% CaO) are invariably produced from bituminous coals and are predominantly composed of aluminosilicate glasses (60 to 90%) with varying amounts of crystalline quartz, mullite, hematite, and magnetite (ACI 232, 2003). These crystalline phases are essentially inert in concrete, and the glass requires a source of alkali or lime (for example, $Ca(OH)_2$) to react and form cementitious hydrates. Such fly ashes are pozzolanic and display no significant hydraulic behavior. High-calcium fly ashes (>20% CaO) may be produced from lignite or subbituminous coals and are comprised of calcium-aluminosilicate glass and a wide variety of crystalline phases, in addition to those found in low-calcium fly ash (see Table 2.4). Some of these crystalline phases will react with water, and this, coupled with the more reactive nature of the calcium-bearing glass, makes these fly ashes react more rapidly than low-calcium fly ashes and renders

Table 2.4 Crystalline phases in fly ash

Mineral name	Chemical formula	Comments
Quartz	SiO_2	These phases are inert in concrete and are the only crystalline phases in low-calcium fly ash.
Mullite	$Al_6Si_2O_3$	
Hematite	Fe_2O_3	
Magnetite	Fe_3O_4	These phases are also present in lesser amounts in high-calcium fly ash.
Anhydrite	$CaSO_4$	
Tricalcium aluminate (C₃A)	$Ca_3Al_2O_6$	Many of these phases react with water to form solid products of hydration. The quantity of these phases generally increases as the calcium content of the fly ash increases. These phases are not found in fly ashes with low to moderate calcium content (<15% CaO).
Dicalcium silicate (C₂S)	Ca_2SiO_4	
Lime	CaO	
Periclase	MgO	
Melilite	$Ca_2(Mg,Al)(Al,Si)_2O_7$	
Merwinite	$Ca_3Mg(SiO_4)_2$	
Alkali sulfates	$(Na,K)_2O_4$	
Sodalite	$Ca_2(Ca,Na)_6(Al,Si)_{12}O_{24}(SO_4)_{1-2}$	

Source: Data from ACI 232, *Use of Fly Ash in Concrete*, ACI Committee 232 Report, ACI 232.2R-03, American Concrete Institute, Farmington Hills, MI, 2003.

the fly ash both pozzolanic and hydraulic in nature. These fly ashes will react and harden when mixed with water due to the formation of cementitious hydration products. Fly ash with intermediate calcium contents (8 to 20% CaO) falls somewhere between low-calcium and high-calcium fly ashes in terms of composition and reactivity. Table 2.4 lists the crystalline phases that may be detected in fly ash.

Figure 2.4 compares x-ray diffraction (XRD) patterns for Class F and Class C fly ash (from Mehta, 1983), slag (Moranville-Regourd, 1998), and silica fume (Schlorholtz, 2006). The difference in the crystalline phases is apparent from the size and location of the sharp peaks. Glass does not produce a well-defined peak in an XRD pattern, but if there is a sufficient quantity of glass present, it will produce a broad hump or diffuse band in the diffractogram. The position of this hump depends on the composition of the glass, and it will be located near the main peak of the crystalline compound if the glass is devitrified (Diamond, 1983a). The data from Mehta (1983) in Figure 2.4 show that the position of the diffuse band is very different for Class F and Class C fly ashes, and Diamond (1983a) has shown that the 2θ angle at which the maximum value of the band occurs increases as the calcium content of the fly ash increases, up to about 20% CaO, as shown in Figure 2.5. Diamond (1983a) ascribes this shift to the changing structure of the aluminosilica glass as the calcium content of the glass increases. There is a sudden shift in the position of the diffuse band for fly ashes with calcium contents above 20%, and very little change in

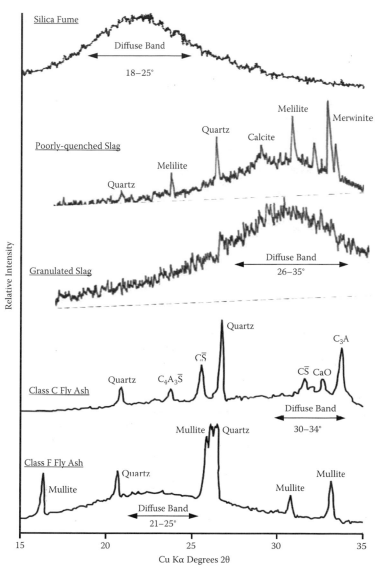

Figure 2.4 X-ray diffraction patterns for Class F and Class C fly ash, granulated and poorly quenched slag, and silica fume. (Data from Mehta, P.K., in *Proceedings of the First International Conference on the Use of Fly Ash, Silica Fume, Slag and Other Mineral By-Products in Concrete*, ACI SP-79, Vol. I, America Concrete Institute, Detroit, 1983, pp. 1–46; Moranville-Regourd, M., in *Lea's Chemistry of Cement and Concrete*, ed. P.C. Hewlett, Arnold, London, 1998, pp. 633–674; Schlorholtz, S., in *Significance of Tests and Properties of Concrete and Concrete-Making Materials*, ed. J.F. Lamond and J.H. Pielert, ASTM STP 169D, American Society of Testing and Materials, West Conshohocken, PA, 2006, pp. 495–511.)

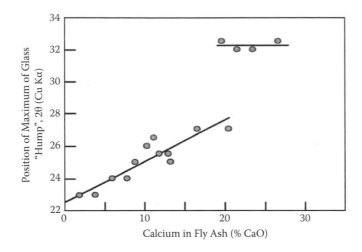

Figure 2.5 Effect of calcium in fly ash on the position of the glass hump in XRD. (From Diamond, S., *Cement and Concrete Research*, 13(4), 459–464, 1983. Reproduced with permission from Elsevier.)

position with increasing calcium above 20%. This sudden change in location has been ascribed to the formation of calcium-aluminate glass in fly ash with more than 20% CaO (Diamond, 1983a). Interestingly, the position of the glass hump for high-calcium fly ash occurs at a similar position as that for the glass in slag (see Figure 2.4).

Figure 2.6 shows data from semiquantitative XRD analysis (McCarthy et al., 1990) revealing the relationship between the C_3A and calcium content of fly ash. Fly ashes with less than about 14% CaO contain no C_3A that is detectable by XRD. As the CaO content of fly ash increases above 14% CaO there is a general trend of increasing C_3A content. It should be noted that XRD may not detect microcrystalline C_3A in fly ash. C_3A is one of the more important crystalline components in fly ash because it contributes to the self-hardening of high-calcium fly ash through the formation of ettringite and influences the sulfate resistance of concrete produced with such fly ash (discussed in Chapter 9). The same study (McCarthy et al., 1990) also showed that both the sulfate content and the free lime content of fly ash increased with the CaO content.

In addition to providing an indication of the mineralogy and reactivity of the fly ash, the calcium content is also useful in predicting how effective the fly ash will be in terms of reducing the heat of hydration (Thomas et al., 1995), controlling expansion due to alkali-silica reaction (Shehata and Thomas, 2000), and providing resistance to sulfate attack (Shashiprakash and Thomas, 2001). These issues are addressed in Chapters 5, 6, and 9.

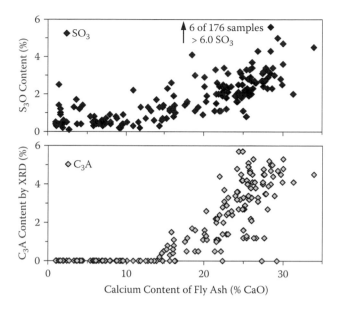

Figure 2.6 Effect of calcium in fly ash on the C_3A content (as determined by XRD) and SO_3 content. (Data from McCarthy, G.J., et al., in *MRS Symposium on Fly Ash and Coal Conversion By-Products Characterization, Utilization and Disposal V*, ed. R.L. Day and F.P. Glasses, Vol. 178, Materials Research Society, Pittsburgh, PA, 1990, pp. 3–33.)

The quantity of alkalis in fly ash can range from less than 1% Na_2Oe[1] up to 10% Na_2Oe (Thomas, 2007). The ratio of sodium to potassium (Na_2O/K_2O) generally increases with increasing CaO content, as does the proportion of the alkalis that are "available" as determined by the available alkali test in ASTC C311 (McCarthy et al., 1990). Most fly ashes have alkali contents below 3% Na_2Oe, but there are a few sources that produce fly ashes with much higher alkali levels (for example, 5 to 10% Na_2Oe). These fly ashes tend to be very reactive, as the alkalis raise the pH of the pore solution when they are mixed in concrete and the high pH accelerates the dissolution of the glass in the fly ash. Particular attention should be paid to the (alkali-silica) reactivity of aggregates when high-alkali fly ashes are used in concrete.

The sulfate content of fly ash generally ranges from less than 0.1 to 5% SO_3, with an average value of around 1.5% SO_3 (Thomas, 2007). There is a tendency for the sulfate content to increase as the calcium content of the fly ash increases (see Figure 2.6) due to the free lime combining with SOx in the flue gases. In exceptional cases the sulfate content may exceed 5% SO_3, and in some sources the sulfate content has exceeded 7% SO_3. Many

---

[1] Equivalent alkali: $Na_2Oe = Na_2O + 0.658 \times K_2O$.

specifications limit the sulfate content of fly ash, as will be discussed in Chapter 10.

Fly ash contains a small amount of unburned carbon. The amount of carbon, usually determined indirectly by measuring the loss on ignition (LOI), depends on a number of factors, including the fineness and composition of the coal, and the burning conditions in the furnace. LOI values tend to be lower in high-calcium fly ash, probably as a result of the more complete burning of the softer coals from which such ashes are derived. The use in concrete of fly ash with high carbon content can lead to increased water demand and problems entraining air. This will be discussed in Chapter 5.

Most fly ash particles are either solid spheres or hollow cenospheres, but some plerospheres, which are open particles filled with smaller spheres, may also be present. The particle size distribution of fly ash varies widely, with many particles being less than 1 μm in diameter and some being more than 100 times larger (see Figure 2.7). The median particle size varies between fly ashes from different sources, but is generally in the range from 5 to 20 μm. As shown in Figure 2.7, fly ash particles are generally in the size range of portland cement particles with a similar median particle size. However,

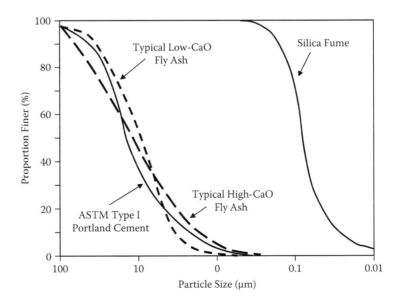

Figure 2.7 Typical particle size distributions for portland cement, fly ash, and silica fume. (From Mehta, P.K., in *Proceedings of the First International Conference on the Use of Fly Ash, Silica Fume, Slag and Other Mineral By-Products in Concrete*, ACI SP-79, Vol. I, America Concrete Institute, Detroit, 1983, pp. 1–46. Reproduced with permission from the American Concrete Institute.)

some fly ashes have a broader range of particle sizes than portland cement and a higher proportion of submicron particles. The spherical shape, broad particle size distribution, and presence of submicron particles all contribute to the ability of fly ash to decrease the water demand of concrete (discussed in Chapter 5).

The specific surface area of fly ash generally lies in the range of 300 to 500 m²/kg (1500 to 2400 ft²/lb), although both lower and higher values may be encountered. The bulk density of fly ash depends on the degree of compaction of the powdered material and may be anywhere from 500 to 1500 kg/m³ (35 to 95 lb/ft³). The specific gravity ranges from as low as 1.9 to as high as 2.8. The value is dependent on the composition of the fly ash, especially the carbon and iron content, although it has also been shown to increase with calcium content due to the formation of a denser glass (McCarthy et al., 1990). The color of fly ash varies from being almost white, through buff colored, to light gray. High carbon contents will tend to produce darker gray colors.

2.2.3 Fly ash beneficiation

In most cases, fly ash is taken straight from the precipitators and used in concrete with no need for material processing. However, sometimes the fly ash produced from a certain source does not meet the requirements for use in concrete, or its performance is less than that required for a certain market, and in such cases, the fly ash may be processed to improve its properties. The process of improving the quality of the fly ash is called beneficiation, and the properties of fly ash that are beneficiated include the fineness and carbon content of the material.

It is well established that increasing the fineness of fly ash can improve its performance in concrete, by reducing water demand, increasing the rate of strength gain at early ages, and improving concrete durability (Butler, 1981; Berry et al., 1989; Obla et al., 2003). The fineness of fly ash can be increased by mechanical or air classification (Hassan and Cabrera, 1998), or by grinding (Bouzoubaa et al., 1997).

The carbon content of a fly ash can be reduced by separating the carbon from the fly ash electrostatically, a process known as triboelectric separation (Bittner and Gasiorowski, 1999), by burning the residual carbon in a fluidized bed (Cochran and Boyd, 1993), or by flotation (Groppo, 2001). Processes for reducing the negative impact of excess carbon without reducing the carbon content of the fly ash have been proposed, and these techniques involve modifying the surface activity of the carbon (Sabanegh et al., 1997).

Table 2.5 Classes of fly ash used in North America

Specification	Class	Description in specification	Chemical requirements
ASTM C618	F	Fly ash normally produced from burning anthracite or bituminous coal Fly ash has pozzolanic properties	$SiO_2 + Al_2O_3 + Fe_2O_3 \geq 70\%$
	C	Fly ash normally produced from lignite or subbituminous coal Has some cementitious properties in addition to pozzolanic properties Some Class C fly ashes may contain lime contents higher than 10%	$SiO_2 + Al_2O_3 + Fe_2O_3 \geq 50\%$
CSA A3001	F	Low-calcium fly ash	$CaO < 15\%$
	CI	Moderate-calcium fly ash	$15\% \leq CaO \geq 20\%$
	CH	High-calcium fly ash	$CaO > 20\%$

2.2.4 Fly ash classification

The most widely used specification for fly ash in North America is ASTM C 618: *Standard Specification for Coal Fly Ash and Raw or Calcined Natural Pozzolan for Use in Concrete (Equivalent to AASHTO M 295).* This specification divides fly ash into two classes based on its source of origin and composition, as shown in Table 2.5. Many fly ashes produced from lignite or subbituminous coals meet the chemical requirement of Class F fly ash ($SiO_2 + Al_2O_3 + Fe_2O_3 \geq 70\%$). Such fly ashes may be classed as Class F or Class C and are sometimes referred to as Class F/C or C/F fly ashes. In Canada, the specification covering fly ash is CSA A3001: *Cementitious Materials for Use in Concrete*, which separates fly ash into three types based on the calcium content of the fly ash, as shown in Table 2.5.

Figure 2.8 shows the distribution in terms of calcium content of fly ashes from 110 commercially available sources in North America (Thomas, 2007). Figure 2.9 compares the calcium content with the sum of the oxides ($SiO_2 + Al_2O_3 + Fe_2O_3$) for the same fly ashes. Most fly ashes that meet CSA Type F or Type CH would be classified as, respectively, Class F and Class C by ASTM C 618.

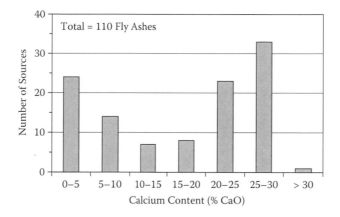

Figure 2.8 Calcium oxide content of 110 commercially available fly ashes from North America. (From Thomas, M.D.A., *Optimizing the Use of Fly Ash in Concrete*, PCA IS548, Portland Cement Association, Skokie, IL, 2007. Reproduced with permission from the Portland Cement Association.)

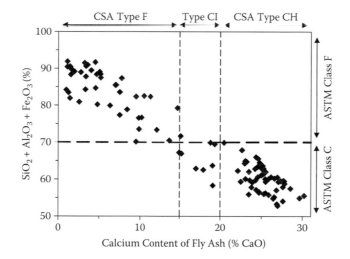

Figure 2.9 Comparison of calcium oxide (CaO) and the sum of oxides (SiO_2 + Al_2O_3 + Fe_2O_3) for 110 commercially available fly ashes from North America. (Modified from Thomas, M.D.A., *Optimizing the Use of Fly Ash in Concrete*, PCA IS548, Portland Cement Association, Skokie, IL, 2007.)

2.3 SLAG

2.3.1 Production of slag

Ground granulated blast furnace slag is the glassy material formed from molten slag produced in blast furnaces as an industrial by-product from the production of iron used in steelmaking. Figure 2.10 shows a schematic of an iron blast furnace. The charge, consisting of iron ore, a fluxing stone (usually a combination of limestone and dolomite), and coke as fuel, is continuously fed into the top of the furnace while hot air is blasted in farther down. As the burden moves down the furnace it ignites and reaches temperatures in excess of 2000°C (3600°F). At this temperature the iron oxide from the ore is reduced to molten (pig) iron, which sinks to the bottom of the furnace due to its high density. The slag is formed from the remaining ingredients, which consist of the calcium and magnesium from the fluxing stone, the alumina and silica gangues from the iron ore, and a small amount of ash from the coke. These ingredients form a fused melt that floats above the denser molten iron. Both the iron and slag are tapped or drawn off from the furnace at regular intervals, and it is a continuous (rather than batch) process. The pig iron goes to make steel and the by-product slag is used in a number of different applications. The material may be poured into trenches and allowed to air cool, being later crushed

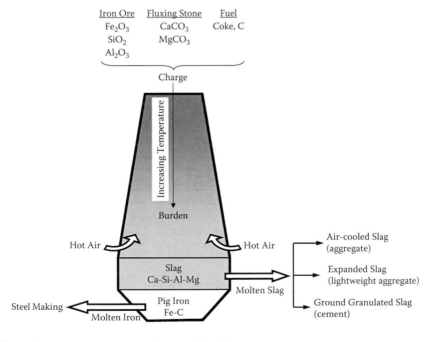

Figure 2.10 Schematic layout of an iron blast furnace.

and used as a normal-density aggregate. Alternatively, expanded slag light-weight aggregate can be produced by mixing the molten slag with water. With appropriate processing the slag can be made suitable for use as a cementitious material for concrete.

To transform the molten slag into a cementing material, it must be rapidly cooled (granulated), dried, and then ground to a fine powder. The resulting product is known as ground granulated (iron) blast furnace slag. This material is commonly referred to by the acronym GGBS in Europe, or as slag cement, or simply slag, in North America. From here on the material will be referred to as slag, meaning only the specially processed material from an iron blast furnace. Slags from other pyrometallurgical processes or from other industries are not discussed in this textbook. Note that steel slag resulting from the steelmaking process, which occurs subsequent to the production of pig iron in an iron blast furnace, is not the same as ground granulated (iron) blast furnace slag and is generally unsuitable for use as a cementing material in concrete.

To produce cementitious slag, the molten slag from the blast furnace is cooled rapidly (quenched) from around 1500°C (2700°F) by quenching with water to form a glassy structure. The process involves pouring the slag in a controlled manner into a stream of water and is termed granulation, as the resulting product has the appearance of coarse granules of sand. Only if the slag is rapidly cooled will it vitrify (form a glassy structure), and it is this glass that displays hydraulic or cementitious properties. If the slag is allowed to cool slowly in air, it will form crystalline products that are basically inert in concrete; in other words, it will have no cementitious value.

The granulated slag is then dried before being ground in a ball mill, or similar equipment, to produce a fine powder material of approximately the same or greater fineness than portland cement.

An alternative to water granulation is the pelletizing process. This process involves passing the molten slag through a series of water jets and then passing it over a spinning drum that breaks the molten stream into a series of pellets that are flung into the air. The process was initially developed to improve the efficiency of handling the waste product, but it was later found that by increasing the amount of water spray and the speed of rotation of the drum, a sufficient amount of vitrification occurs to produce the desired glassy structure. Less water is required for pelletization than granulation, and with sufficient control the need for drying prior to grinding is eliminated. Furthermore, pelletized slag requires less processing energy.

2.3.2 Composition and properties of slag

As with fly ash, the performance of slag is strongly influenced by its physical, chemical, and mineralogical composition. Slag manufacturers have little control over the chemical composition of the slag, as this is dictated

Table 2.6 Range of chemical compositions encountered with slags from the United States and Canada

Oxide	Range (%)
SiO_2	32–42
Al_2O_3	7–16
CaO	32–45
MgO	5–15
S	0.7–2.2
Fe_2O_3	0.1–1.5
MnO	0.2–1.0

Source: ACI 233, *Slag Cement in Concrete and Mortar*, ACI Committee 233 Report, ACI233R-03, American Concrete Institute, Farmington Hills, MI, 2003. Printed with permission from the American Concrete Institute.

by the iron-making process. The slag is composed of the noncombustible and nonferrous components of the raw ingredients that are used to charge the blast furnace. A typical chemical analysis of a North American slag is compared with other SCMs in Table 2.2.[1] These data together with the composition shown on a ternary plot in Figure 2.2 show that the composition of slag is closer to that of portland cement than those of other SCMs, although slag is much lower in calcium and higher in silica than portland cement. Table 2.6 shows the range of compositions encountered with slags from the United States and Canada (ACI 233, 2003) and these data show that the composition of slag can vary in terms of its major components (CaO, Al_2O_3, SiO_2, and MgO). However, because iron production is a highly controlled process, the composition of the waste stream (slag) from a particular source is very consistent compared to fly ash.

Figure 2.4 compares XRD patterns for rapidly quenched and poorly quenched slags (Moranville-Regourd, 1998). If rapidly quenched, little crystalline product is formed and the slag is entirely composed of glass, which produces a broad hump rather than sharp peaks when analyzed by XRD. If the molten material is cooled more slowly, crystalline products such as merwinite, melilite, calcite, and quartz will form, producing sharp peaks in the XRD pattern. The presence of glass is considered the most important factor in determining the hydraulicity of slag, as crystalline slag possesses little significant reactivity (Hooton, 1987). Although some investigators have demonstrated that there is a linear relationship between the glass content of slag and the strength of mortars, other researchers (Smolczyk, 1980; Demoulian et al., 1980) have shown that small quantities of crystalline material (up to 5%) have been found to increase strength, and

---

[1] It should be noted that although the residual iron present in slag is frequently reported as ferric oxide (Fe_2O_3), it is present as ferrous sulfide, FeS.

that the presence of up to 30% crystalline merwinite had little significant impact on strength. Furthermore, it is difficult to quantify the glass content of slag, and the methods available for its determination tend to produce different results (Hooton, 1987).

The chemical composition of slag also affects its reactivity. Generally the reactivity increases with increasing alkalinity and with increasing calcium, alumina, and magnesium content, at least up 10% MgO (Odler, 2000), and decreasing silica and magnesia contents (Hooton, 1987). The early strength of mortars or concretes tends to increase as the alumina content of the slag increases, and this is considered to result from an increase in the amount of ettringite that forms at early ages (Odler, 2000). A number of hydraulic moduli (HM), based on the basicity (ratio of basic to acidic oxides), have been proposed to predict the reactivity of the slag, the most widely used modulus being that determined by Equation 2.1:

$$HM = \frac{CaO + MgO + Al_2O_3}{SiO_2} \qquad (2.1)$$

Some national specifications have incorporated minimum values of HM $\geq$ 1.0 (Germany) or 1.4 (Japan), whereas the British Standard (BS 6699) has a minimum value of HM $\geq$ 1.0, but does not include the alumina content in the calculation of the modulus.

The reactivity of the slag also increases with increasing fineness. Generally, slag is ground to a fineness in the range of 400 to 650 m^2/kg, which produces angular-shaped particles with a similar particle size distribution as portland cement (Figure 2.1).

The bulk density is around 1200 kg/m^3 (75 lb/ft^3), and the specific gravity averages 2.9. The finished product resembles white cement.

2.3.3 Classification of slag

In North America, slag is classified on the basis of its slag activity index (SAI), which is determined on the basis of the strength of mortars in accordance with Equation 2.2.

$$SAI = \frac{SP}{P} \times 100\% \qquad (2.2)$$

where SP and P are the compressive strengths of mortar cubes with, respectively, 50 and 0% (control) slag. ASTM C989 provides for three grades of slag based on the 7-day and 28-day strength activity indices, as shown in Table 2.7.

Table 2.7 Strength grades of slag in ASTM C989

		Slag activity index, minimum%	
	Grade	Avg. last 5 tests	Any single test
7-day index	80	—	—
	100	75	70
	120	95	90
28-day index	80	75	70
	100	95	90
	120	115	110

2.4 SILICA FUME

2.4.1 Production of silica fume

Silica fume is the ultrafine noncrystalline silica produced in electric arc furnaces as an industrial by-product of the production of silicon metals and ferrosilicon alloys. Silica fume is also known as condensed silica fume or microsilica. A schematic of the process is shown in Figure 2.11. In silicon metal production, a source of high-purity silica (such as quartz or quartzite), together with wood chips and coal, is heated to around 1800°C (3300°F) in an electric arc furnace to remove the oxygen from the silica (reducing conditions). On heating the silica (SiO_2) and carbon (C) together, most of the SiO_2 is reduced to silicon metal (Si), which is periodically tapped from the furnace. The remaining silica is only partially reduced to SiO, and this is carried away with carbon monoxide (CO) by the exhaust gases. As these gases are drawn away from the furnace, the SiO oxidizes and returns to its original state of SiO_2. As the temperature drops in the flue the silica condenses into small droplets of silica glass, which are removed from the exhaust gases by a series of filter bags. The collected product is referred to as silica fume (or microsilica).

2.4.2 Composition and properties of silica fume

The silica content of silica fume is very high compared to those of other SCMs used in concrete (see Table 2.2 and Figure 2.2) and varies with the type of alloy being produced. Ferrosilicon alloys are produced with varying amounts of silicon (50 to 98% Si); alloys with more than 98% silicon (Si) are referred to as silicon metal. As the silicon (Si) content of the alloy product increases, so does the silica (SiO_2) content of the by-product silica fume, as shown in Table 2.8. Many national standards (United States, Norway, France, Japan, and Australia) and the European standard covering silica fume for use in concrete require that a minimum silica content of SiO_2 be

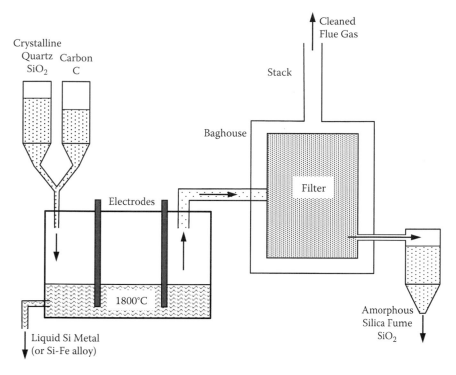

Figure 2.11 Schematic of a silicon metal/alloy furnace and baghouse. (Modified from Fidjestol, P., and Lewis, R., in *Lea's Chemistry of Cement and Concrete*, ed. P.C. Hewlett, Arnold, London, 1998, pp. 675–708.)

Table 2.8 Typical silica contents of silica fume collected from the production of different ferrosilicon alloys

Alloy	Typical silica content of silica fume (% SiO_2)
50% ferrosilicon	74–84
75% ferrosilicon	84–91
Silicon metal (≥98% Si)	87–98

Source: Data from ACI 234, *Guide for the Use of Silica Fume in Concrete*, ACI Committee 234 Report, ACI 234R-06, American Concrete Institute, Farmington Hills, MI, 2006.

≥85%. In Canada, two grades of silica fume are recognized, Types SF1 and SF2, with minimum silica contents of 75 and 85% SiO_2, respectively. Experience with silica fume in concrete is largely restricted to the by-products from ferrosilicon alloys with 75% silicon (Si) or higher. Silica fume from other silicon alloys is also produced, but such material may not be suitable for use in concrete (ACI 234, 2006).

As the silica content of silica fume decreases, the amounts of iron, alumina, and alkalis tend to increase. Most silica fumes currently used in concrete contain more than 90% SiO_2.

X-ray diffraction patterns (Figure 2.4) reveal silica fume to be essentially amorphous (noncrystalline) with a broad hump centered around the location of the peak for cristobalite (ACI 234, 2006). Trace amounts of crystalline compounds, such as silicon carbide and quartz, or silicon metal may be present in silica fume.

Silica fume consists of extremely fine, spherical, glassy particles (see Figures 2.1 and 2.12) with an average diameter of 0.1 to 0.2 µm. Table 2.1 compares the properties of silica fume with those of other SCMs. The average size of individual silica fume particles is approximately 100 times smaller than those of portland cement or other SCMs. The specific surface area of silica fume is in the range of 15,000 to 25,000 m²/kg (measured by nitrogen absorption) compared with values of 300 to 600 m²/kg for other cementing materials. However, the surface area of silica fume is not measured by the Blaine apparatus typically used for other cementing materials, and the values are not directly comparable. The specific gravity of silica fume is approximately 2.2, but may be higher (up to 2.5) if the iron content is high. The color of as-produced silica fume is dark gray to black, although the material can be processed to a white color. The bulk density of the as-produced material is in the range of 130 to 430 kg/m³, but this is frequently increased by mechanically densifying the material, as discussed in the next section.

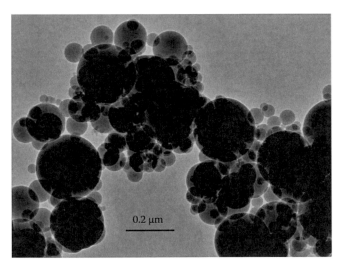

Figure 2.12 Transmission electron microscope image of silica fume. (Micrograph courtesy of Elkem AS.)

2.4.3 Silica fume product forms

Because of its extremely fine particle size and low bulk density, as-produced silica fume is difficult to handle. It is more usually available as a densified or slurried product. Densification may be achieved by passing compressed air through a silo containing silica fume, causing the particles to tumble and collide, resulting in loosely bound agglomerations. Alternatively, densification can be achieved by mechanical means, such as passing the material through a screw auger. The bulk density is increased to above 500 kg/m³ by compaction. The agglomerated particles are readily dispersed by the mixing action in mortar or concrete unless the product is overdensified. Water-based slurries typically contain around 50% by mass of silica fume, with resulting densities around 1400 kg/m³. Some slurried products may contain chemical admixtures. Silica fume may also be pelletized by mixing the undensified product with small amounts of water to form hard pellets of various sizes; pelletizing aids the disposal of silica fume (e.g., in landfills). Pelletized silica fume is unsuitable for use as a separately added constituent of concrete, as the pelletized particles will not be broken up and dispersed in the concrete mixer. However, pelletized silica fume is interground with a portland cement clinker to produce blended cements. Table 2.9 summarizes the different product forms of silica fume.

Table 2.9 Silica fume product forms

Form	Bulk density (kg/m³)	Description
Undensified	130–430	As produced.
Densified (or compacted)	500–600	Fine particles are agglomerated by tumbling action in compressed air or by mechanical compaction. Agglomerations readily break up due to shearing action in mortar or concrete mixers.
Slurried	1320–1440	Water-based slurry produced typically with 40 to 60% silica fume by mass. Proprietary products containing chemical admixtures (e.g., high-range water reducers and set retarders) may be available.
Pelletized	Varies	Hard pellets formed by mixing silica fume with small quantities of water. Pellets may be landfilled or interground with portland cement clinker to produce blended cement, but should *not* be used directly as a constituent of concrete.

2.5 NATURAL POZZOLANS

2.5.1 Types of natural pozzolans

A natural pozzolan is defined by the American Concrete Institute (ACI 116, 2000) as "either a raw or calcined natural material that has pozzolanic properties (for example, volcanic ash or pumicite, opaline chert and shales, tuffs and some diatomaceous earths)." As such, natural pozzolans include a wide variety of materials with a broad range of properties (Table 2.10 and Figure 2.13). As mentioned in Chapter 1, natural pozzolans such as volcanic ash were used together with lime to produce the first hydraulic binders more than 2000 years ago. Although they have been used for millennia, and are currently still used, they are less commonly used in most countries today than readily available industrial by-products such as fly ash and slag.

It should be noted that an alternate definition of natural pozzolan is materials that "do not require any further treatment apart from grinding," thus differentiating such materials from artificial pozzolans that require "chemical and/or structural modifications of materials having no or only weak pozzolanic properties" (Massazza, 1998). This definition would classify volcanic ashes and tuffs as natural pozzolans, whereas calcined or thermally treated materials (such as clays, shales, and diatomaceous earths) would be classed together with fly ash and silica fume as artificial pozzolans. While

Table 2.10 Chemical composition of some natural pozzolans

Pozzolan	Composition (%)						
	SiO_2	Al_2O_3	Fe_2O_3	CaO	MgO	Na_2Oe	LOI
Roman tuff (Italy)	44.7	18.9	10.1	10.3	4.4	6.7	4.4
Santorin earth (Greece)	65.1	14.5	5.5	3.0	1.1	6.5	3.5
Rhenish trass (Germany)	53.0	16.0	6.0	7.0	3.0	6.0	9.0
Jalisco pumice (Mexico)	68.7	14.8	2.3	—	0.5	9.3	5.6
Diatomaceous earth (United States)	86.0	2.3	1.8	—	0.6	0.4	5.2
Rice husk ash	92.15	0.41	0.21	0.41	0.45	2.39	2.77
Metakaolin	51.52	40.18	1.23	2.00	0.12	0.53	2.01
Moler (Denmark)	75.6	8.62	6.72	1.10	1.34	1.36	2.15
Gaize (France)	79.55	7.10	3.20	2.40	1.04	—	5.90
Opaline Shale	65.4	10.1	4.2	4.6	2.7	1.4	6.3

Source: Data from Mehta, P.K., in *Supplementary Cementing Materials for Concrete*, ed. V.M. Malhotra, CANMET-SP-86-8E, Canadian Government Publishing Center, Supply and Services, Ottawa, Canada, 1987, pp. 1–33; Malhotra, V.M., and Mehta, P.K., *Pozzolanic and Cementitious Materials, Advances in Concrete Technology*, Vol. 1, Gordon and Breach Publishers, Amsterdam, 1996; Massazza, F., in *Lea's Chemistry of Cement and Concrete*, ed. P.C. Hewlett, Arnold, London, 1998, pp. 471–631.

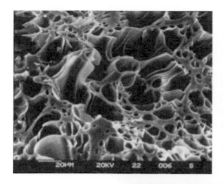
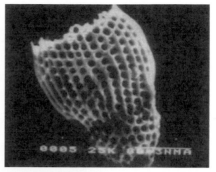
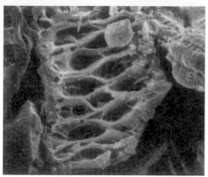

Figure 2.13 SEM images of volcanic pumice ash (top left), diatomaceous earth (top right), and rice husk ash (bottom). (From Malhotra, V.M., and Mehta, P.K., *Pozzolanic and Cementitious Materials, Advances in Concrete Technology*, Vol. 1, Gordon and Breach Publishers, Amsterdam, 1996.)

this classification has merit, for the purposes of this text the term *natural pozzolans* will include calcined and thermally treated natural materials.

A brief description of some of the more commonly used natural pozzolans is included here. For more information the reader is directed to Massazza (1998), ACI (ACI 232, 2012), and Mehta (1987).

2.5.2 Glassy volcanic pozzolans

Volcanic ashes derive their pozzolanic reactivity from the presence of volcanic glass, which forms during the rapid cooling of ash from explosive volcanic eruptions. The most well known of these materials are Santorin earth, which was produced by the volcanic eruption on the island of Santorini (Thera) in Greece around 1600 to 1500 B.C., and Pozzolana, which was produced during the eruption of Mount Vesuvius in 79 A.D. (ACI 232, 2012). Although the chemical composition of volcanic ashes varies between and within sources, the main reactive component is aluminosilicate glass. The chemical reactivity is enhanced by the high surface area of the glass, which has a microporous structure (see Figure 2.13) due to the release of gases in the liquid magma (Massazza, 1998; Mehta, 1987).

2.5.3 Volcanic tuff

The weathering of volcanic glass can alter the glass into zeolite or clay minerals (Massazza, 1998). Zeolitization generally improves the pozzolanic activity, whereas argillation (transformation into clay minerals) reduces pozzolanic activity, requiring such minerals to be calcined or thermally activated to improve pozzolanicity. Rhenish trass is an example of a zeolitic tuff that has been used in pozzolan-lime mortars for about 2000 years and can be found in Roman structures along the Rhine in Germany. It is still used today in combination with portland cement, and there is a German standard covering the material for use in concrete.

2.5.4 Calcined clay and shales

As mentioned above, the argillation of volcanic glasses produces clay minerals with little or no pozzolanic activity. However, clay minerals undergo chemical and structural alterations when heated, with the loss of combined water leading to destruction of the crystalline network. This can leave silica and alumina in an unstable and amorphous state, thereby improving the pozzolanic properties. The resulting pozzolanic properties will depend on the content and composition of the clay minerals and the nature of the thermal treatment (calcining conditions). Typically thermal treatment or calcining is carried out in a rotary kiln using temperatures in the range of 600 to 1000°C.

Metakaolin is produced by the thermal activation of kaolin clay and is a pozzolanic material that has been used in concrete since the construction of a series of large dams in Brazil beginning in 1962 (Saad et al., 1982). High-reactivity metakaolin (HRM) is a term that has been applied to thermally activated high-purity kaolin clay (also known as China clay) to distinguish it from less reactive calcined mixed clays (Gruber et al., 2001). Metakaolin is generally heated to a temperature in the range of 650 to 800°C, which is generally lower than the calcining temperature for other clays and shales. The material is ground to a very fine particle size (1 to 2 μm) to increase its reactivity. Some sources of metakaolin have a high Hunter L whiteness value of 90 or more (on a scale of 0 for black to 100 for white), which makes them suitable for architectural concrete applications. The physical and chemical properties are compared to those of other SCMs in Tables 2.1 and 2.2, and other natural pozzolans in Table 2.10.

2.5.5 Diatomaceous earth and opaline silica

Diatomaceous earths are sedimentary rocks that consist of hydrous amorphous silica that is formed from the siliceous skeletons of aquatic microorganisms (diatoms). Diatomaceous earths are highly pozzolanic but are usually mixed with clay minerals and invariably require calcining to make them suitable for use in concrete. According to Massazza (1998), the largest deposit of diatomites occurs in California. Calcined diatomaceous shales and clays were used in the construction of a number of large concrete structures by the U.S. Bureau of Reclamation in the 1950s (Elfert, 1974; ACI 232, 2012). Other sources of opaline silica that have been widely used as pozzolanic materials include gaize from France and moler from Denmark, both of which have been used "as is" or after calcination.

2.5.6 Rice husk ash

Rice husk ash (RHA) is produced from the incineration of rice husks. After combustion of the cellulose and lignin components of the rice husks, the residual ash is predominantly amorphous silica with a cellular structure (Table 2.10 and Figure 2.13). The pozzolanic properties of RHA and its behavior in concrete are highly dependent on combustion conditions. Mehta (1992) has shown that, under carefully controlled conditions, it is possible to make a highly reactive pozzolan suitable for use in concrete. However, this resource has not yet been exploited as a supplementary cementing material.

Chapter 3

Chemical reactions of SCMs in concrete

3.1 POZZOLANIC REACTIONS

By definition, a pozzolan is "a siliceous or siliceous and aluminous material ... that reacts chemically with calcium hydroxide (lime) to form compounds having cementitious properties." In early civilizations, natural pozzolans were mixed together with lime to produce hydraulic cements, whereas today natural or artificial pozzolans are usually used together with portland cement. The hydration of the calcium silicate compounds, C_3S (or alite) and C_2S (or belite), in portland cement produces abundant calcium hydroxide through the reactions represented by Equations 3.1 and 3.2 (see Table 3.1 for cement chemistry nomenclature).

$$2C_3S + 11H \rightarrow C_3S_2H_8 + 3CH \tag{3.1}$$

$$2C_2S + 9H \rightarrow C_3S_2H_8 + CH \tag{3.2}$$

The composition of the calcium silicate gel that forms from the hydration of the alite and belite varies. The actual calcium-to-silicon atomic ratio is typically greater than the value of 1.5 represented by $C_3S_2H_8$ and may vary up to 2 or more. Because of the variable composition this phase is often simply referred to as calcium silicate hydrate or C-S-H.[1] The C-S-H is the principal cementing compound in portland cement concrete and is largely responsible for providing strength and other properties to the concrete. In addition to C-S-H and CH hydrated portland cement contains aluminoferrite phases (AFm and AFt) produced by the hydration of the other portland

---

[1] Equations 3.1 and 3.2 show the calcium silicate hydrate that forms to have eight molecules of water in its structure. Some references on cement chemistry show only three molecules of water in the structure, that is, $C_3S_2H_3$. Both are correct. $C_3S_2H_8$ includes water that is both chemically combined in the crystal structure and physically bound to the solid surfaces, whereas $C_3S_2H_3$ does not include the physically bound water.

Table 3.1 Cement chemistry nomenclature

Abbreviation	Formula
S	SiO_2
A	Al_2O_3
C	CaO
F	Fe_2O_3
M	MgO
N	Na_2O
K	K_2O
$\overline{S}$	SO_3
$\overline{C}$	CO_2
H	H_2O

cement clinker compounds, C_3A and C_4AF, in the presence of gypsum, $C\overline{S}H_2$.

The reaction of the calcium hydroxide (CH) with the silica (S) component of a pozzolan can be represented by Equation 3.3 (Helmuth, 1987).

$$xCH + yS + zH \rightarrow C_xS_yH_z \qquad (3.3)$$

Typically, the C/S ratio (x/y in Equation 3.3) of the C-S-H that forms from this reaction will be lower than the C/S ratio measured for C-S-H in hydrated portland cement without pozzolan, and the difference will vary depending on the age, type, and amount of pozzolan. For example, Taylor (1997) suggests that the C/S of the C-S-H in 3-day-old pastes containing fly ash is not significantly different than that formed in pure portland cement pastes, but that the ratio decreases with age and with increasing fly ash content. For mature pastes (91-day) with 28% low-CaO fly ash, Taylor (1997) shows the average C/S ratio to be 1.63 compared to a value close to 2.0 for plain portland cement paste. Uchikawa (1986) reports values as low as C/S = 1.01 for the C-S-H in 4-year-old pastes with 40% fly ash. Traetteberg (1978) showed similar values of C/S = 1.1 for pastes with silica fume after periods of just 7 to 14 days.

The alumina present in pozzolans will also react with the CH from portland cement and may produce a variety of phases, the principal ones including strätlingite or gehlenite hydrate (C_2ASH_8) and hydrogarnet (C_3AH_6), and others being calcium aluminate hydrate (C_4AH_{13}), ettringite ($C_3A \cdot 3C\overline{S} \cdot H_{32}$), calcium monosulfoaluminate ($C_3A \cdot C\overline{S} \cdot H_{12}$), and calcium carboaluminate ($C_3A \cdot C\overline{C} \cdot H_{12}$).

The reaction of metakaolin with calcium hydroxide produces mainly C-S-H and C_2ASH_8 according to Equation 3.4 (Helmuth, 1987) with possibly small quantities of C_4AH_{13} (Massazza, 1998).

$$AS_2 + 3CH + zH \rightarrow CSH_{z-5} + C_2ASH_8 \qquad (3.4)$$

The reactions represented in Equations 3.2 and 3.4 ignore the role of the alkali hydroxides. The first stage in the pozzolanic reaction is an attack by the OH^- ions on the silica and aluminosilica framework in the glass and subsequent dissolution. Because of the strong concentration of Na^+ and K^+ in solution, the initial product is likely to be an amorphous alkali-silicate. However, the abundance of calcium in the system and the low solubility of C-S-H mean that the alkali-silicate phase is short-lived.

The pozzolanic activity of a material defines the ability of that material to react with calcium hydroxide. There are two components to this activity, the first being the total amount of CH with which the material can combine and the second being the rate at which the reaction with CH occurs. Massazza (1998) states that there is general agreement that the total amount of CH with which a pozzolan can combine is dependent on the following factors:

- Nature of the reactive phases in the pozzolan
- Content of these phases
- SiO_2 content of these phases
- CH/pozzolan ratio of the mix
- Duration of curing

On the other hand, the rate of the reaction with CH will depend upon:

- Specific surface area of the pozzolan
- Water/solid ratio of the mix
- Temperature

When pozzolans are mixed with portland cement, the rate of reaction will also depend on the composition of the portland cement, in particular its alkali content.

Figure 3.1 shows data from various studies that have measured the CH content of pastes or mortars with various amounts of pozzolan present and illustrates the differences in the pozzolanic activity between pozzolans. The pozzolan with the highest activity among those shown in Figure 3.1 appears to be a high-reactivity metakaolin produced from high-purity china clay from the United Kingdom (Kostuch et al., 1993). A replacement rate of just 20% metakaolin is sufficient to consume all of the CH produced by the portland cement in a period of just 28 days. Silica fume also reacts quickly, but

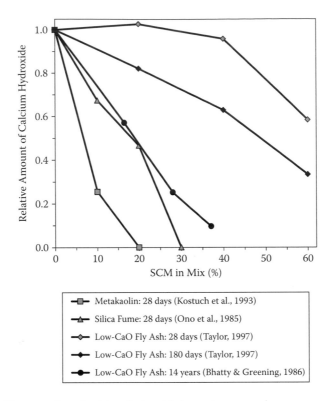

Figure 3.1 Consumption of calcium hydroxide by various pozzolans.

appears to combine with less CH, and a replacement level of 30% is required to consume all of the CH. Fly ash (low CaO in this case) is a much slower-reacting pozzolan, with little significant CH consumption being observed after 28 days. In fact, in this case the use of 20% fly ash actually leads to very small increase in the amount of CH produced by the portland cement in 28 days due to the acceleration of alite hydration (Taylor, 1997). After 180 days, the fly ash consumes significantly more CH. Also shown on the graph are data from a different study with fly ash (Bhatty and Greening, 1986), which shows very significant consumption of CH in mature (14-year-old) fly ash pastes.

The effect of temperature on the rate of CH consumption by fly ash can be seen in Figure 3.2 from the work of Buttler and Walker (1982). The temperature of curing does not significantly change the ultimate amount of CH consumed by the pozzolanic reaction, but increasing the temperature dramatically increases the rate of the reaction. This is also the case for pozzolans other than fly ash. The rate of the pozzolanic reaction is generally more sensitive to changes in temperature than is the hydration of portland cement.

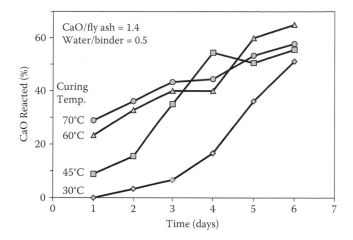

Figure 3.2 Effect of curing temperature on rate of pozzolanic reaction of fly ash. (Data from Buttler, F.G., and Walker, E.J., in *Proceedings of the Use of PFA in Concrete*, Leeds, April, Vol. I, pp. 71–81, 1982.)

Figure 3.3 compares the rate of hydration of fly ash with that of C_3S and slag. There is very little reaction of fly ash during the first few days, and only 10% of the fly ash has reacted by 14 days. There is an increase in the rate of the pozzolanic reaction beyond this time, but approximately half the material is still unreacted at 1 year.

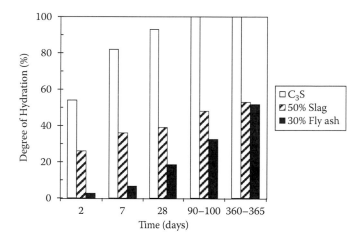

Figure 3.3 Degree of hydration of C_3S in cement clinker, slag, and fly ash in blended cements. (Data from Locher, F.W., *Cement: Principles of Production and Use*, Verlag Bau+Technik GmbH, Düsseldorf, 2006.)

High-CaO fly ashes exhibit both pozzolanic and hydraulic behavior. When mixed with water some high-calcium fly ashes will harden due to the hydration of crystalline calcium silicate, the reaction between free lime and some of the glass, and the formation of ettringite. The hydration products in blends of portland cement and high CaO are essentially the same as those that form in portland cement mixes with the addition of strätlingite and C_2AH_8 (Taylor, 1997), and increased quantities of calcium aluminate hydrates.

3.2 HYDRATION OF SLAG

Ground granulated blast furnace slag is not a pozzolan but is considered to be a latent hydraulic material. If slag is mixed with water it will hydrate, but the hydration products form a thin Si-rich layer on the surface of the slag grains, which essentially stifles any further hydration. An alkaline activator has to be used to raise the pH (>12) in the vicinity of the slag and prevent the formation of an impermeable layer, thereby allowing the continued dissolution of the glass. Suitable activators include calcium hydroxide, sodium or potassium hydroxide, sodium carbonate, or sodium silicate. The hydration of portland cement produces both CH and alkali hydroxides, and portland cement is an excellent activator for the hydration of slag.

The hydration products formed in a mixture of portland cement and slag are similar to those that form in a pure portland cement mix; that is, C-S-H, CH, AFm, and AFt are the main products. However, the quantity of CH that is formed is lower than the amount that would be produced by the hydration of the portland cement, indicating that CH is at least partly consumed by the hydration of the slag. This is shown in Figure 3.4 for pastes made with varying slag contents. The data shown include amounts calculated based on the amount of C_3S that had reacted, assuming no consumption by the slag, and measured amounts. In all cases, there is a significant reduction in the amount of CH beyond what would be expected if the slag did not consume CH. The C/S ratio of the C-S-H is decreased, and the Al_2O_3 and MgO contents are increased when slag is present, the effect becoming more pronounced at higher slag contents and as the hydration process advances. The C-S-H that forms directly from the hydration of the slag (inner product of slag) has a C/S ratio in the range of 1.6 in pastes with a slag content of 40% (Taylor, 1997), compared to about 1.8 for C-S-H outside of the slag grains. For slags high in MgO a hydrotalcite-like phase may form.

Figure 3.3 shows the amount of slag hydrated as a function of time. Clearly the rate of slag hydration is substantially slower than that of portland cement, although it is faster than fly ash at early ages. There is still approximately 50% unreacted slag remaining after 1 year.

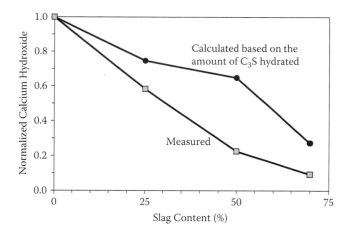

Figure 3.4 Calcium hydroxide concretes in pastes containing varying levels of slag. (Data from Hinrichs, W., and Odler, I., *Advances in Cement Research*, 2, 9–13, 15–20, 1989.)

The rate of reaction of slag when blended with portland cement will depend on the following factors:

- The amount of glass in the slag
- The chemical composition of the slag (basicity)
- The proportion of slag
- The composition of the cement (particularly the alkali content)
- Temperature

3.3 EFFECT OF SCMS ON THE HYDRATION OF PORTLAND CEMENT

The use of supplementary cementing materials (SCMs) can accelerate the early-age hydration of alite when blended with portland cement. Isothermal calorimetry has been used in a number of studies to characterize the influence of pozzolans and slag on the early hydration. Figure 3.5 shows a series of heat development curves for blends of C_3S and different pozzolans. Heat development curves such as these are usually composed of five main stages, as follows:

- **Stage 1:** A period of rapid heat evolution due to the *initial hydrolysis* of the cement. This stage starts as soon as the cement and water come into contact and ceases after about 15 minutes. It is hard to capture the heat evolved during this stage, and the heat evolution for this stage is not shown in Figure 3.5.

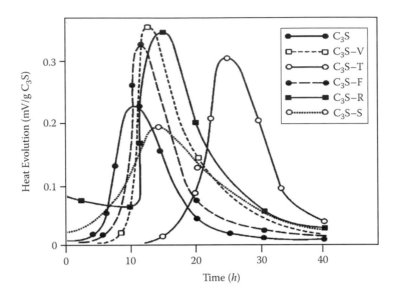

Figure 3.5 Rates of heat evolution in blends of C_3S with various pozzolans: C_3S-T is a blend of C_3S and fly ash. (From Ogawa, K., et al., *Cement and Concrete Research*, 10(5), 683–696, 1980. Printed with permission of Elsevier.)

- **Stage 2:** A period of relative inactivity known as the *dormant period* or the *induction period*. This period results from the need for the ions in solution to reach a critical concentration before nucleation occurs and hydration begins. The duration of this period is typical between 2 and 4 hours, but may be altered by the presence of constituents that act as either set accelerators or retarders. This stage represents the period when fresh concrete can be handled, placed, and consolidated.
- **Stage 3:** The hydration of C_3S begins again and accelerates, reaching a maximum at the end of the *acceleration period*.
- **Stage 4:** The rate of hydration decreases during the *deceleration period* due to the formation of C-S-H around the unhydrated C_3S grains, which acts as a barrier.
- **Stage 5:** The hydration reaches a *steady state* and future hydration is diffusion controlled, as the mass transport of water and dissolved ions through the hydration barrier controls future hydration. The rate of hydration slows as the barrier thickens and approaches completion asymptotically.

Figure 3.5 shows the end of the dormant period, the acceleration and deceleration of the C_3S, and the onset of steady state. Most of the natural pozzolans do not alter the length of the dormant period to any great extent but significantly accelerate the early-age hydration of the C_3S, as shown by an increase in the height of the main peak. This is partly due to the increase in the water-to-C_3S ratio when part of the C_3S is replaced by a pozzolan, but it is mainly due to the finely divided pozzolan providing nucleation sites for the hydrates to form.

One of the pozzolans used in this study was a low-CaO fly ash, and this is shown as C_3S-T in Figure 3.5. It appears that although fly ash still accelerates the hydration of C_3S, it may cause a significant lengthening of the dormant period, which could be manifested as delayed setting in concrete. The reason for this delay is not entirely clear. Studies (Kokubu, 1969) have shown that washing the fly ash with water prior to mixing it with C_3S can eliminate this effect, indicating that the retardation by the unwashed fly ash may be due to some, possibly organic, constituent that is removed during washing.

Using calorimeters to measure the rate of heat evolution in blends of cement and pozzolan, with and without chemical admixtures, is a useful tool for determining the compatibility and predicting setting behavior of material combinations.

3.4 EFFECT OF SCMS ON THE PORE SOLUTION COMPOSITION

The hardened cement paste in portland cement concrete typically has a porosity in the region of 20 to 30%, depending on the initial W/CM and the degree of hydration. In saturated concrete, these pores are filled with a solution of ions and the solution is in equilibrium with the solid hydrates. The development of techniques for extracting the pore solution from hardened cement paste samples (and even mortar and concrete) in the 1970s (Longuet et al., 1973; Barneyback and Diamond, 1981) has permitted studies to be made on the evolution of the pore solution chemistry and factors that affect it, including the use of SCMs. Understanding pore solution effects is particularly important in the study of alkali-aggregate reactions (AARs), as will be discussed in Chapter 9, and other processes, such as corrosion and mass transport.

Figure 3.6 shows the composition of the pore solution for paste and mortar produced with a high-alkali cement (0.91% Na_2Oe) and how it changes with time from an age of approximately 20 minutes to over 18 months; these figures were produced from data presented by Diamond (1983b). The plot on the left-hand side of Figure 3.6 shows the pore solution of a hardened cement paste with W/CM = 0.50 during the first 24 hours of hydration. On mixing cement with water, the alkali sulfates rapidly dissolve and immediately dominate the pore solution together with hydroxyl ions and lesser amounts of calcium. After a few hours, the sulfate concentration starts to decrease due to the formation of ettringite. As the sulfate anions are removed, the hydroxyl ion concentration increases to maintain

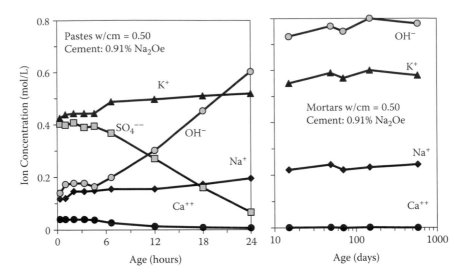

Figure 3.6 Evolution of the pore solution in portland cement paste and mortar. (Data from Diamond, S., in *Proceedings of 6th International Conference on Alkalis in Concrete*, ed. G.M. Idorn and S. Rostam, Danish Concrete Association, Copenhagen, 1983, pp. 155–166.)

electroneutrality with the alkali cations. This increases the pH of the pore solution, which reduces the solubility of calcium hydroxide, and hence the concentration of calcium ions in solution. At 24 hours, the pore solution is dominated by the alkali hydroxides (Na⁺, K⁺, and OH⁻) and has very low concentrations of calcium and still decreasing concentrations of sulfate. The right-hand plot in Figure 3.6 shows the concentration of ions in the pore solution of mortars produced with the same W/CM and high-alkali cement, but with addition of Ottawa quartz sand. Shortly after 1 day, only sodium, potassium, and hydroxyl ions can be detected in the pore solution at significant concentrations, and the pore solution composition remains relatively stable with time beyond this point.

For pastes composed of portland cement with no SCMs, the composition of the pore solution can be predicted from the alkali content of the cement. Figure 3.7 shows that the hydroxyl ion concentration of pastes (*w/c* ~ 0.5) between 28 and 90 days old can be estimated from the alkali content of the portland cement using the relationship in Equation 3.5:

$$OH^- = 0.72 \times PC_{alk} \qquad (3.5)$$

where OH^- is the hydroxyl ion concentration in mMol/L and PC_{alk} is the alkali content of the cement expressed as percent Na_2Oe. The ion concen-

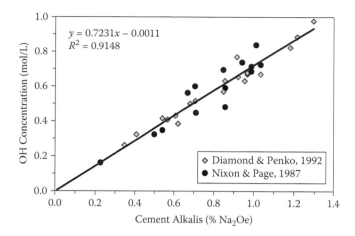

Figure 3.7 Relationship between the hydroxyl ion concentration of the pore solution and the alkali content of the portland cement. (Data from Diamond, S., and Penko, M., in *Durability of Concrete—G.M. Idorn International Symposium*, ed. Jens Holm, ACI SP-131, American Concrete Institute, Detroit, 1992, pp. 153–168; Nixon, P.J., and Page, C.L., in *Concrete Durability, Katherine and Bryant Mather International Conference*, ed. J.M. Scanlon, ACI SP-100, Vol. 2, American Concrete Institute, Detroit, 1987, pp. 1833–1862.)

tration would be expected to be higher or lower, respectively, in pastes of lower or higher *w/c*.

When SCMs are used in concrete, not only do they change the composition of the hydrate phases, but they also impact the chemistry of the pore solution. SCMs, in particular some fly ashes, may contain alkalis in quantities in excess of that normally found in portland cement (0.2 to 1.2% Na_2Oe). However, most SCMs usually bring about a reduction in the concentration of alkalis in the pore solution, and this is attributed to the increased ability of the hydration products that form in the presence of SCMs to bind alkalis. The increased binding ability is thought to result from the generally lower C/S ratio of the C-S-H that forms in concrete containing SCMs.

Figure 3.8 shows pore solution data for a series of studies conducted at the author's laboratory. While most SCMs reduce the alkalinity of the pore solution compared to pastes produced with only high-alkali portland cement, the following general trends are observed from these data and data from other studies (Thomas, 2011):

- The influence of the SCM on the pore solution depends on the composition and amount of SCM used, and the maturity of the paste.
- SCMs with high surface area bring about a more rapid consumption of alkalis.

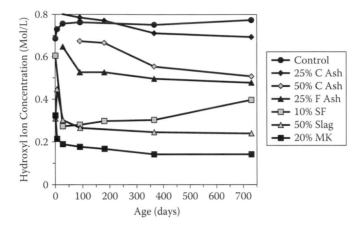

Figure 3.8 Pore solution evolution for pastes with high-alkali cement and various SCMs. (Data from Shehata, M.H., et al., *Cement and Concrete Research*, 29, 1915–1920, 1999; Bleszynski, R., The Performance and Durability of Concrete with Ternary Blends of Silica Fume and Blastfurnace Slag, PhD thesis, University of Toronto, 2002; Ramlochan, T., and Thomas, M.D.A., in *Proceedings of the 4th ACI/CANMET International Conference on the Durability of Concrete*, ed. V.M. Malhotra, ACI SP-192, Vol. I, pp. 239–251, 2000.)

- SCMs high in reactive silica and low in calcium and alkali are the most effective in terms of lower pore solution alkalinity.
- High-calcium fly ash is less effective at reducing the pore solution alkalinity than that with low-calcium content.

Generally the concentration of alkalis in the pore solution of a blended cement paste decreases as the alkali content of the binder decreases, obviously, but also as the calcium content decreases and the silica content increases. Figure 3.9 shows the relationship between the hydroxyl ion concentration in the pore solution of 90-day-old paste samples (W/CM = 0.50) and a chemical parameter based on the Na_2Oe, CaO, and SiO_2 content of the binder (including the portland cement and all SCMs present). The data in Figure 3.9 represent 79 different blends of cementing materials, including portland cements with a range of alkali contents (0.35 to 1.09% Na_2Oe), fly ashes with a range of calcium and alkali contents (1.1 to 30.0% CaO and 1.4 to 9.6% Na_2Oe), silica fume, metakaolin, and slag; control mixes were cast with portland cement only, binary mixes with cement plus one SCM, and ternary mixes with cement plus two SCMs.

The chemical parameter used in Figure 3.9 is $(Na_2Oe \times CaO)/(SiO_2)^2$, and the contents of soda, lime, and silica used in the calculation are based

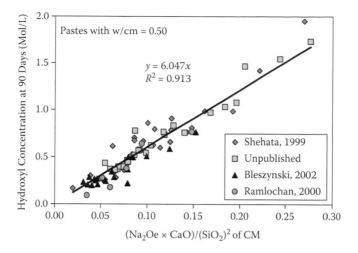

Figure 3.9 Relationship between the alkalinity of the pore solution and the chemical composition of the binder. (From Thomas, M.D.A., *Cement and Concrete Research*, 41, 1224–1231, 2011.)

on all of the components in the binder. It is not proposed that the relationship in Figure 3.9 be used to predict the hydroxyl ion concentration of the pore solution; rather it is presented to illustrate the importance of the constituents in determining the availability of the alkali in the system.

The alkali content of the pore solution plays a major role in determining the risk of damage due to alkali-silica reaction, as discussed in Chapter 9.

Chapter 4

Microstructure of portland cement—SCM systems

4.1 PORE STRUCTURE

Freshly mixed cement paste consists of cement particles suspended in water. As the cement hydrates, forming calcium-silicate hydrate (C-S-H), calcium hydroxide (CH), and various aluminoferrite phases (AFm and AFt), the space originally filled with water is reduced because the volume of the hydrates is greater than the volume of the unhydrated cement. Figure 4.1 shows how the relative volumes of the major compounds change as a portland cement paste (W/CM = 0.50) hydrates and fills up the space originally occupied by water.[1] The remaining (originally water-filled) space is termed capillary porosity because it is generally considered that the pores making up this space are of a size in which capillary effects can occur. The capillary porosity clearly reduces as the curing period increases, and it will also reduce as the W/CM is decreased, because with lower W/CM there is less water-filled space to begin with (Figure 4.2).

As the capillary porosity is reduced, the strength of the paste increases and its permeability decreases (Figure 4.3), illustrating the importance of curing and W/CM on the quality of concrete. As hydration proceeds and the volume of capillary pores decreases, the size of pores decreases and the capillary pore system becomes less well connected and more tortuous. These changes in the structure of the pore system are referred to as pore structure refinements and render the porous material (hardened cement paste) less permeable and more resistant to the penetration of fluids and deleterious agents. The sizes of pores in hydrated cement paste vary widely from capillary pores, which may be anywhere in the size range from 10 μm down to 10 nm, to tiny gel pores, which are generally considered to

---

[1] For W/CM = 0.50, water initially occupies 61% of the volume of the paste, the remaining 39% being occupied by unhydrated cement. According to measurements made by Powers and Brownyard (1948; Powers 1961), fully hydrated portland cement occupies just over twice (2.14 times) the volume of the unhydrated cement. Consequently, the volume of the originally water-filled space will reduce to just 17% when a paste with W/CM = 0.50 is completely hydrated.

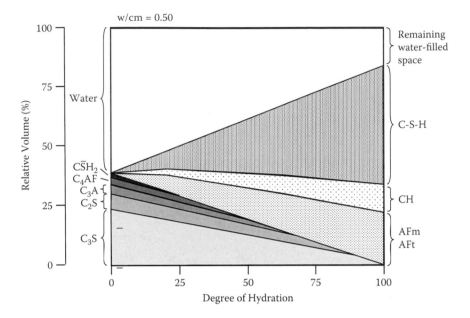

Figure 4.1 Relative volumes of the major compounds as a function of the degree of hydration estimated by a computer model for W/CM = 0.50. (From Tennis, P.D., and Jennings, H.M., *Cement and Concrete Research*, 855–863, 2000. Printed with permission from the Portland Cement Association.)

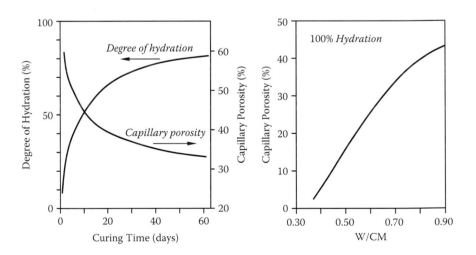

Figure 4.2 Relationships between capillary porosity, curing time, degree of hydration, and W/CM calculated from Paver's equations. (From Young, J.F., et al., *The Science and Technology of Civil Engineering Materials*, Prentice Hall, Upper Saddle River, NJ, 1998. Printed with permission from Prentice Hall.)

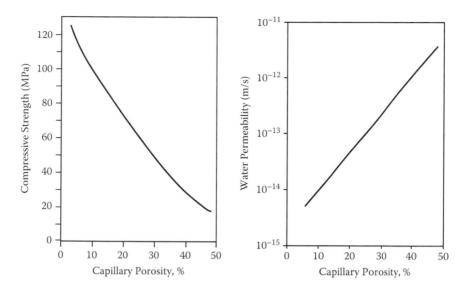

Figure 4.3 Relationship between strength and capillary porosity. (Modified from Young, J.F., et al., *The Science and Technology of Civil Engineering Materials*, Prentice Hall, Upper Saddle River, NJ, 1998; permeability converted to a log scale. Printed with permission from Prentice Hall.)

range from 10 nm down to less than 0.5 nm and are an intrinsic part of the C-S-H gel (Mindess et al., 2003). Table 4.1 provides a classification for the different pores.

Porosity measurements are commonly made by oven drying a saturated sample and converting the mass of evaporable water to the volume voids in the sample. This procedure removes the water from all pores and provides a measure of the total porosity contributed by the capillary pores plus the gel pores. As such, it is not really a very useful measurement, as two materials can have the same total porosity, but very different pore structures, and hence mass transport properties. Information on the pore size distribution can be obtained from a number of techniques, and the two that are commonly used in the study of cement and concrete are mercury intrusion porosimetry (MIP) and gas absorption. Mercury intrusion is useful for characterizing the capillary pore system, whereas gas absorption is better suited for measuring the finer pores.

Figure 4.4 shows MIP data for a portland cement paste at ages 3 and 14 days. It is possible to calculate the actual pore size and volume of pores within a particular size range using MIP, but this involves making a number of assumptions that are considered to be invalid for cement paste (Diamond, 2000); consequently, MIP results are only used for comparative purposes in this text. The data in Figure 4.4 show that both extending the

Table 4.1 Classification of pores sizes and influence on properties of hydrated cement paste

Designation	Size range	Description	Role of water	Properties affected
Capillary pores	10,000–50 nm (10–0.05 μm)	Macropores (large capillaries)	Behaves as bulk water	Largely responsible for mass transport (permeability and diffusion)
	50–10 nm	Medium mesopores (medium capillaries)	Small surface tension forces generated	Mass transport in absence of macropores Responsible for shrinkage above 80% RH
	10–2.5 nm	Small mesopores	Large surface tension forces generated	Shrinkage between 80 and 50% RH
Gel pores	2.5–0.5 nm ≤0.5 nm	Micropores Interlayer spaces between Ca-Si sheets in C-S-H	Strongly absorbed water; no menisci form	Shrinkage at all RH Creep

Source: Modified from Mindess, S., et al., *Concrete*, 2nd ed., Prentice Hall, Englewood Cliffs, NJ, 2003. Printed with permission from Prentice Hall.

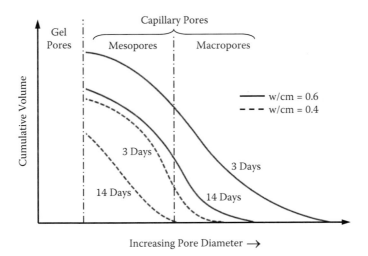

Figure 4.4 Effect of age and W/CM on pore size distributions in portland cement paste. (From Young, J.F., et al., *The Science and Technology of Civil Engineering Materials*, Prentice Hall, Upper Saddle River, NJ, 1998. Printed with permission from Prentice Hall.)

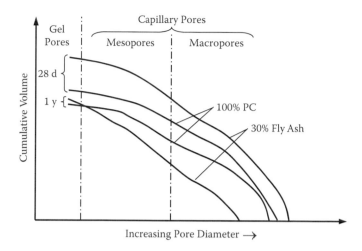

Figure 4.5 Effect of fly ash on pore size distribution at 28 days and 1 year. (Data from Manmohan, D., and Mehta, P.K., *Cement, Concrete and Aggregates*, 3(1), 63–67, 1981.)

moist curing period and reducing the W/CM have two effects on the pore size distributions. First, the total volume of capillary pores is reduced, and second, there is a shift of the curve to the left indicating a refinement to the pore structure. The volume of the larger macropores is greatly reduced, whereas the volume of the smaller mesopores is less affected. This presumably indicates that the large pores are being filled in with hydrates containing gel pores that are too fine to be intruded by mercury.

The use of SCMs in cement and concrete has a profound effect on the pore size distribution. Figure 4.5 shows the effect of low-calcium fly ash on the pore size distributions at 28 days and 1 year for pastes with W/CM = 0.50 (from Manmohan and Mehta, 1981). At 28 days, the partial replacement of cement with 30% fly ash results in a higher porosity and a greater volume of larger pores (macropores). Extending the curing to 1 year reduces the volume of the pores measured in both pastes, but the difference is most marked for the fly ash. There is a very significant reduction in the capillary-sized pores between 28 days and 1 year for the fly ash paste, and within this classification of pores there is a significant shift from the macropores to mesopores. At 1 year, the pore structure of the fly ash paste is more refined than that of the plain cement paste (100% portland cement (PC)). The improvements in pore structure with time emphasize the importance of curing, especially with slowly reacting pozzolans, if the potential benefits associated with their use are to be realized.

Figure 4.6 shows the effect of silica fume at more moderate levels of replacement (from Sellevold and Nilsen, 1987). Even a replacement level

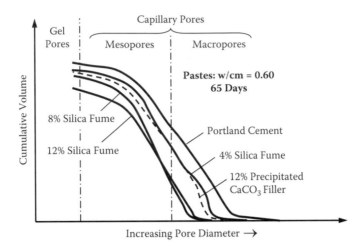

Figure 4.6 Effect of silica fume and precipitated CaCO₃ filler on pore size distribu-
tion in hardened cement paste. (Data from Sellevold, E.J., and Nilsen, T., in
Supplementary Cementing Materials for Concrete, ed. V.M. Malhotra, CANMET,
Ottawa, 1987, pp. 165–243.)

of 4% silica fume has a significant impact on the pore size distribution,
reducing the volume of the macropores. This study included an inert filler
(precipitated calcium carbonate) of apparently similar fineness as the silica
fume, which when used at a replacement level of 12% did improve the pore
structure somewhat, but not to the same extent as the same level of silica
fume, indicating that the pozzolanic reaction is largely responsible for the
pore structure refinement. The incorporation of silica fume did not significantly change the total porosity of theses pastes, as measured by water
absorption, emphasizing the importance of measuring the pore size distribution rather than the total volume of pores. Sellevold and Nilsen (1987)
also reported that cement could be replaced by silica fume on a 3:1 basis (3
parts cement replaced by 1 part silica fume) without increasing the capillary porosity, indicating silica fume to be three times as efficient as cement
in this regard. Cementing efficiency factors for supplementary cementing
materials are discussed in Chapter 7.

Figure 4.7 shows data for mortars containing 20% metakaolin and
pastes containing 40% slag. The use of these materials to partially replace
portland cement also results in significant refinements to the pore structure
in terms of a reduced volume of macropores.

Because of concerns regarding the validity of mercury intrusion porosimetry for measuring the pore size distributions in hydrated cement systems
(Diamond, 2000), other techniques, such as solvent exchange and absorption, have been used to study the pore size distributions in cement paste samples. Thomas (1989) used methanol absorption to study the effect of fly ash

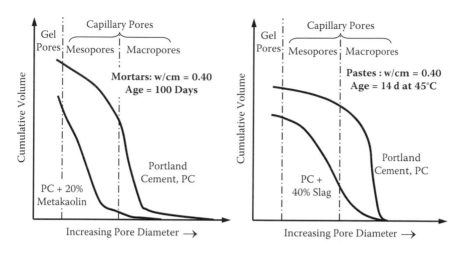

Figure 4.7 Effect of metakaolin and slag on the pore size distribution of hardened cement pastes. (Data from Kostuch, J.A., et al., in *Proceedings of Concrete 2000*, ed. R.K. Dhir and M.R. Jones, Vol. 2, University of Dundee, Scotland, 1993, pp. 1799–1811; Roy, D.M., and Parker, K.M., in *Proceedings of the 1st International Conference on the Use of Fly Ash, Silica Fume, Slag and Other Mineral By-Products in Concrete*, ed. V.M. Malhotra, SP-79, Vol. 1, American Concrete Institute, Farmington Hills, MI, 1983, pp. 397–414.)

and curing on the pore structure. In this technique, hydrated cement paste samples were immersed in methanol for 14 days prior to drying *in vacuo* (to minimize damage to the pore structure caused by removing water), and the dried samples were then placed in an absorption vessel with methanol and two reference glasses of single pore diameters of 4 and 37 (36.8) nm. By weighing the samples during stepwise absorption and assuming that pores of equal size in the cement paste and reference materials become filled simultaneously, it is possible to calculate the volume of pores in the cement paste with diameters smaller than 4 and 37 nm. Total porosity is determined by oven drying the samples once they are saturated with methanol at the end of the test. Figure 4.8 shows some of the data from the study by Thomas (1989). The two graphs in the top half of the figure show how the total volume of pores and the volume of pores <37 nm and <4 nm change with time in plain portland cement paste and paste with 30% fly ash. The total porosity changes by only a small amount after the first measurement made at 1 day. However, there is a significant increase in the volume of pores <4 nm during curing, and this is presumably due to the production of C-S-H gel and associated gel pores. The volume of pores <37 nm also increases, although this increase is largely due to the increase in the volume of the sub-4-nm pores. Concomitant with the production of gel is the significant reduction in the pores larger than 37 nm, as represented by the difference

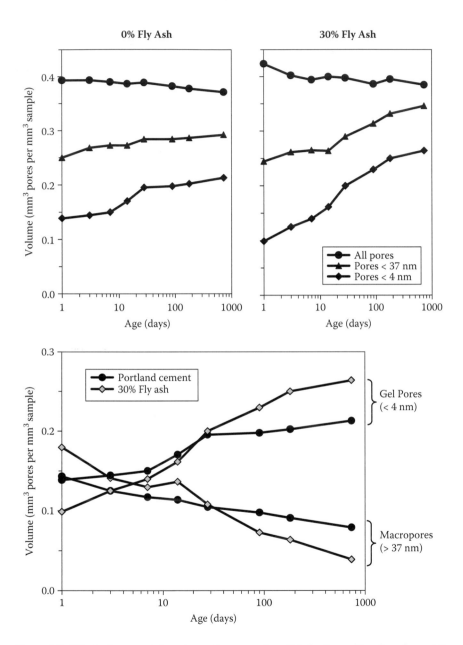

Figure 4.8 Effect of age and fly ash on the pore size distribution of hardened cement paste as measured by the solvent absorption technique. (Data from Thomas, M.D.A., *Advances in Cement Research*, 2(8), 181–188, 1989.)

between the lines representing the total porosity and the pores <37 nm. The graph in the bottom half of Figure 2.8 compares the increase in gel pores (<4 nm) and the decrease in macropores (>37 nm) for the two different cement paste samples. The effect of fly ash is very evident. At 1 day there is less gel porosity and more macroporosity in the fly ash paste, but beyond 7 days the converse is the case. After an extended period of curing (2 years) the macro-porosity of the fly ash paste is slightly less than half of that measured in the portland cement paste. The same study (Thomas, 1989) also demonstrated the importance of curing, as the pore structure refinements observed for well-cured fly ash pastes did not occur in pastes when the relative humidity of the storage environment was below 80%.

4.2 INTERFACIAL TRANSITION ZONE (ITZ)

The cement paste adjacent to aggregate particles tends to be weaker and more porous than the bulk cement paste because the small cement grains are not able to pack densely close to the larger aggregate surface; this is illustrated in Figure 4.9. This weaker region is commonly referred to as the interfacial transition zone (ITZ), as the properties are inferior to the bulk paste. The ITZ is characterized by a higher porosity and higher W/CM than the bulk paste, and there is a tendency for large crystals of calcium

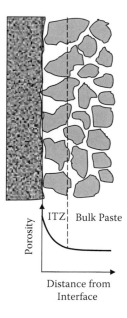

Figure 4.9 Interfacial transition zone (ITZ) forms due to poor packing of cement grains close to the aggregate surface.

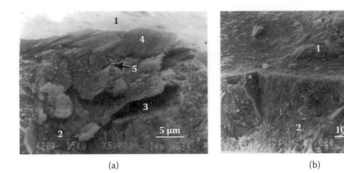

(a) (b)

Figure 4.10 ITZ in cement paste in 28-day-old concrete without (left) and with (right) silica fume. Legend: 1 = aggregate, 2 = cement paste, 3 = voids, 4 = calcium hydroxide, 5 = microcracks. (From Bentur, A., and Cohen, M.D., *Journal of the American Ceramic Society*, 70(10), 738–743, 1987; ACI 234, *Guide for the Use of Silica Fume in Concrete*, ACI Committee 234 Report, ACI 234R-06, American Concrete Institute, Farmington Hills, MI, 2006. Printed with permission from the American Concrete Institute.)

hydroxide (CH) and ettringite to form in this zone. SCMs, particularly finely divided pozzolans such as silica fume, have been shown to improve (or densify) the ITZ by improving the particle packing at the surface and by converting some of the CH that forms there to C-S-H. Figure 4.10 shows images from scanning electron microscopy that illustrate the effect of silica fume (from Bentur and Cohen, 1987). In the concrete without silica fume voids, microcracks and deposits of CH are visible in the ITZ between the dense C-S-H and the aggregate, whereas in the concrete with silica fume there does not appear to be a transition zone between the aggregate and the C-S-H.

A good illustration of the ITZ and the strengthening effects of SCMs is to compare the fracture surface of high-strength concrete with and without SCM after testing for compressive strength. In the absence of SCM, fracture will occur around aggregate particles as the ITZ is the "weak link" and the surface will comprise unbroken aggregate particles. However, if the paste is strong enough and SCMs are present, fracture often occurs through aggregate particles, as the ITZ is stronger than the aggregate.

4.3 PORE BLOCKING AND MASS TRANSPORT

The total porosity of pastes containing SCMs is not significantly reduced compared to a portland cement paste of the same W/CM and maturity, but the pore structure refinements discussed above result in an increased volume of fine pores and decreased volume of capillary pores. As a consequence, the connectivity of the capillary pore structure is reduced in pastes containing SCMs, and this has a significant impact on the mass transport properties of the hydrated paste. The effect of pozzolans and slag on the mass transport properties of concrete will be discussed in detail in Chapter 9 on durability.

Chapter 5

Properties of fresh concrete

5.1 WORKABILITY AND WATER DEMAND

The term *workability* refers to the ease with which fresh concrete can be mixed, placed, molded, consolidated, and finished. Similarly, the term *consistency* applies to the ability of the freshly mixed concrete to flow. Although there are a wide variety of test methods for measuring these properties, by far the most widely used is the slump test, which measures the consistency of the concrete in terms of the subsidence (to the nearest ¼ in. or 5 mm) of a molded specimen immediately after removal of the mold (a truncated cone).

The water demand of concrete refers to the amount of water required to achieve the desired level of workability or consistency (slump). The water demand of a portland cement concrete is largely controlled by the size, shape, surface texture, and grading of the aggregates (particularly the coarse aggregate), but is also a function of the amount and fineness of the cement, and the quantity of entrained air present. Workability and consistency are also affected by temperature.

Water-reducing admixtures and superplasticizers can be used to significantly increase the workability and consistency, or reduce the water demand of concrete, and are widely used in concrete production today. Supplementary cementing materials can also have a significant impact on the workability, consistency, and water demand of the concrete. The effect depends on the type of supplementary cementing material (SCM) and the level of replacement, among other factors.

The use of good quality fly ash with a high fineness and low carbon content improves the workability and reduces the water demand of concrete, and consequently, the use of fly ash should permit the concrete to be produced at a lower water content than a portland cement concrete of the same workability (Figures 5.1 and 5.2). Although the exact amount of water reduction varies widely with the nature of the fly ash and other parameters of the mix, an approximation is that each 10% of fly ash should allow a water reduction of at least 3%.

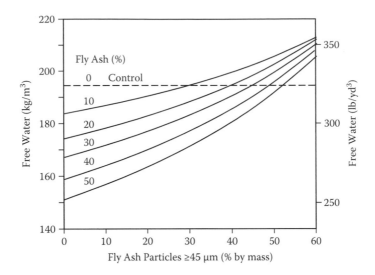

Figure 5.1 Effect of fly ash fineness on water demand of concretes proportioned for equal slump. (Data from Owens, P.L., *Concrete Magazine*, July 1979, pp. 22–26.)

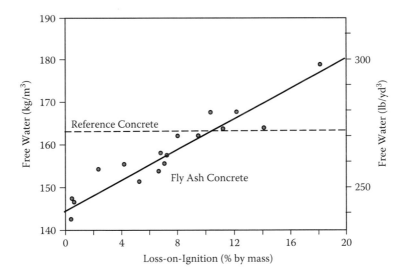

Figure 5.2 Effect of fly ash loss-on-ignition (LOI) on water demand of concretes proportioned for equal slump. (From Sturrup, V.R., et al., in *Proceedings of the 1st International Conference on the Use of Fly Ash, Silica Fume, Slag and Other Mineral By-Products in Concrete*, ed. V.M. Malhotra, ACI SP-79, Vol. 1, American Concrete Institute, Farmington Hills, MI, 1983, pp. 71–86. Printed with permission from the American Concrete Institute.)

Finer fly ashes tend to produce greater reductions in water demand, and fly ashes with a significant 45-μm (#325 sieve) residue may even increase water demand, as shown in Figure 5.1 (Owens, 1979). The effect of the coarse particles can be explained by particle interference, as they tend to be irregular rather than spherical in shape. Grinding coarse fly ash separately or intergrinding it with portland cement clinker improves its ability to reduce the water demand of concrete (Monk, 1973). Fly ashes that have high LOI contents also tend to be less effective in reducing the water demand of concrete (see Figure 5.2), and this is attributed to the absorption of "free water" by unburned carbon particles.

The effect of fly ash on the workability of concrete has been variously attributed to the following phenomena:

- The so-called ball-bearing effect: The spherical particles and smooth surface of fly ash act to lubricate the mix, reducing interparticle friction and increasing flow. Although this explanation is the most widely used and easiest to visualize, it is considered by some to be overly simplistic (Helmuth, 1987). However, it does serve to explain why fly ash is a much more effective water reducer than other pozzolans and slag of similar particle size but irregular shape.
- Increased paste volume: In mixes where cement is replaced with fly ash on an equal mass basis due to the lower specific gravity of fly ash.
- Adsorption-dispersion: The glassy submicron particles of fly ash, having essentially the same surface charge, adsorb onto the surfaces of cement grains with opposite charge, thereby acting as a dispersing agent (see Figure 5.3).

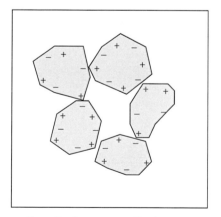 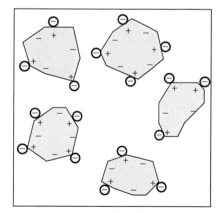

Opposite charges on portland cement particles cause flocculation

Negatively charged fly ash neutralizes positive charges on cement resulting in dispersion

Figure 5.3 Submicron fly ash particles adsorb on and disperse portland cement particles.

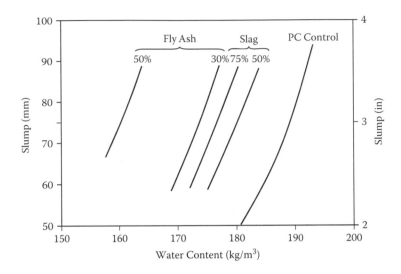

Figure 5.4 Effect of fly ash and slag on slump and water demand. (Data from Concrete Society, *The Use of GGBS and PFA in Concrete*, Technical Report 40, Concrete Society, Wexham, Slough, 1991.)

- The broad particle size distribution of fly ash and particle shape leads to increased particle packing of the cementing materials.

Slag improves the workability and reduces the water demand of concrete, but to a lesser extent than high-fineness fly ash with low carbon content. Figure 5.4 compares the relationship between water content and slump for concrete mixes with varying levels of fly ash and slag, but without water-reducing admixtures. The reduced impact of slag, compared with fly ash, can be explained on the basis of its irregular particle shape and possible lower proportion of similarly charged submicron particles to produce the adsorption-dispersion effect. For a given level of slump, water reductions of between 3 and 5% can be achieved in slag concrete, compared with port-land cement concrete.

At relatively low levels of replacement (<3 to 4%), silica fume may reduce the water demand of concrete as the small spherical particles lubricate the mix and increase particle packing by filling the space between the larger cement grains that would otherwise be filled with water. However, the high surface area of silica fume tends to increase water demand, and this effect dominates as the level of replacement increases above about 5%. In concrete with moderate to high levels (8 to 16%) of silica fume Sellevold and Radjy (1983) estimated that an additional 1 kg of water is required for every 1 kg of silica fume (1 lb of water for every 1 lb of silica fume), as shown in Figure 5.5. Water-reducing admixtures or superplasticizers

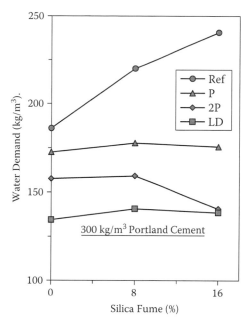

Figure 5.5 in the chart area contains the following labels and annotations:

- Y-axis: Water Demand (kg/m³), scale 100, 150, 200, 250
- X-axis: Silica Fume (%), scale 0, 8, 16
- Legend: Ref, P, 2P, LD
- 300 kg/m³ Portland Cement

Water required to maintain slump = 120 to 130 mm

Ref: Concrete without water-reducing admixture

P: Lignosulfate (40% solution) added at 1 kg solution per 10 kg silica fume

2P: Double the dose for P above

LD: Powdered naphthalene sulfonate added at 1.5 kg powder per 10 kg silica fume

Note: Mixes with 0% silica fume contained same amount of admixture as the mix in the same series with 8% silica fume

Figure 5.5 Effect of silica fume on water demand of concrete (slump = 120 to 130 mm) with and without water-reducing admixtures. (Data from Sellevold, E.J., and Radjy, F.F., in *Proceedings of the 1st International Conference on the Use of Fly Ash, Silica Fume, Slag and Other Mineral By-Products in Concrete*, ed. V.M. Malhotra, ACI SP-79, Vol. 2, American Concrete Institute, Detroit, 1983, pp. 677–694.)

should always be used in concrete containing more than 4 to 5% silica fume by mass of cementing material to both offset the increase in the water demand and ensure adequate dispersion of the silica fume throughout the concrete mix. Figure 5.5 shows that water demand in concrete containing high-range water reducers (lignosulfate or naphthalene sulfonate) is similar for concrete without silica fume and concrete with 8% silica fume. Note that Figure 5.5 indicates a reduced water demand in the presence of high-range water-reducing admixtures (HRWRAs) when 16% silica fume is present in the concrete, but this is misleading, as these concretes contained twice the amount of HRWRA than the mixes with 0 or 8% silica fume.

The effect of natural pozzolans on the workability and water demand varies widely, depending mainly on the physical properties (size, shape, and surface texture) of the material. Many natural pozzolans are porous and can increase the water demand due to the water absorbed by the poz-zolan. However, the absorbed water does not increase the porosity of the concrete and may be available for hydration once the concrete has hard-ened. Metakaolin has little effect on the water demand at low levels of replacement (~5%) but, like silica fume, requires the use of water-reducing or superplasticizing admixtures at higher levels of replacement.

Well-proportioned concrete mixtures containing SCM will have improved workability when compared with a portland cement concrete of the same slump. This means that, at a given slump, concrete with SCM flows and consolidates better than a conventional portland cement concrete when vibrated. This is especially noticeable for concrete containing silica fume or concrete with high levels of fly ash and slag. Such concretes are sometimes described as thixotropic.

The use of SCM also improves the cohesiveness and reduces segregation of concrete. These effects are especially apparent in concrete with moderate levels of silica fume (5 to 10%) or relatively high levels of fly ash (>30%) or slag (>50%), and this allows higher slump concrete to be used without segregation. The spherical particle shape of fly ash also lubricates the mix, rendering it easier to pump and reducing wear on equipment (Best and Lane, 1980).

It should be emphasized that these benefits will only be realized in well-proportioned concrete. The fresh properties of concrete are strongly influenced by the mixture proportions, including the type and amount of cementing material, the water content, the grading of the aggregate, the presence of entrained air, and the use of chemical admixtures.

5.2 BLEEDING

Bleeding is defined as the upward migration of mixing water in fresh concrete caused by the settlement of the solid materials and results in the development of a layer of water at the surface of newly placed concrete. Bleeding is normal, and provided it does not occur excessively, it is not detrimental to the concrete and may be helpful in reducing plastic shrinkage cracking. Excessive bleeding occurs when the ratio of the volume of water to the surface area of solids is high and may diminish the quality of the concrete due to (1) the accumulation of bleed water under aggregate particles or embedded steel weakening the bond, (2) an increase in the W/CM at the surface, especially if the finishing takes place when the bleed water is present, (3) an accumulation of bleed water and development of a weak layer below the surface if the surface is finished before bleeding has ceased, (4) the development of bleed channels that may present preferred pathways for the ingress of aggressive species, and (5) excessive settlement of the solid concrete. Excessive bleeding may occur in improperly proportioned concrete and is exacerbated by high mixing water contents, low cement contents, and the use of aggregates deficient in fines. Bleeding is lower in air-entrained concrete.

In properly proportioned concrete containing moderate levels of fly ash bleeding is reduced primarily due to the reduced water content. If advantage is not taken of the reduced water demand when fly ash is used, then the amount of bleeding may increase. The slower setting of concrete containing fly ash usually means that bleeding will occur over a longer period, and this will extend the time until final finishing operations can take place.

The effect of slag on bleeding depends to a large extent on the fineness of the slag. If the slag is ground finer than portland cement, it will decrease the rate of bleeding, and conversely, coarser slag increases bleeding. If the slag is used at a high replacement level and its use allows for a reduction in the water content, then bleeding may be reduced significantly.

The use of silica fume in concrete with low to moderate W/CM (≤ 0.45) almost eliminates bleeding due to the need to wet the high surface area of the silica fume particles and due to the presence of an increased volume of submicron particles, which tend to block bleed channels.

Concrete with normal levels of silica fume or relatively high levels of fly ash and slag, that exhibits little bleeding, should be finished as quickly as possible and immediately protected to prevent plastic shrinkage cracking when the ambient conditions are such that rapid evaporation of surface moisture is likely. The guidance given in ACI 305, *Hot Weather Concreting*, should be followed.

5.3 AIR ENTRAINMENT

Concrete containing low-calcium (Class F) fly ashes generally requires a higher dose of air-entraining admixture to achieve a satisfactory air-void system. This is mainly due to the presence of unburned carbon, which absorbs the admixture. Consequently, higher doses of air-entraining admixture are required as either the fly ash content of the concrete increases or the carbon content of the fly ash increases. Figure 5.6 shows the relationship between the carbon content in fly ash concrete and the amount of air-entraining admixture (AEA) required to produce an air content of $6.5 \pm 1\%$ (Sturrup et al., 1983). The carbon content of the fly ash used to generate the data in Figure 5.6 shows unusually high variability. The carbon content of fly ash is usually measured indirectly by determining its LOI. The increased demand for air-entraining admixture should not present a significant problem to the concrete producer provided the carbon content of the fly ash does not vary significantly between deliveries. It has been shown that as the admixture dose required for a specific air content increases, the rate of air loss also increases (Gebler and Klieger, 1983).

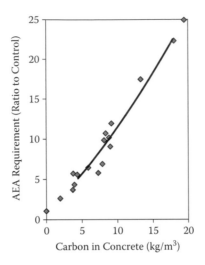

Figure 5.6 Effect of carbon from fly ash on the air-entraining admixture (AEA) required for concrete to achieve 6.5 ± 1% air. (From Sturrup, V.R., et al., in *Proceedings of the 1st International Conference on the Use of Fly Ash, Silica Fume, Slag and Other Mineral By-Products in Concrete*, ed. V.M. Malhotra, ACI SP-79, Vol. 1, American Concrete Institute, Farmington Hills, MI, 1983, pp. 71–86. Printed with permission from the American Concrete Institute.)

Generally, high-calcium fly ashes require a smaller increase in the air entrainment dose than low-calcium fly ashes. Some Class C fly ashes high in water-soluble alkali may even require less admixture than those mixes without fly ash (Pistilli, 1983).

The use of silica fume, slag, and natural pozzolans can also affect the air-entraining dose due to increases in the surface area of the cementing materials component of the concrete mixture. For slag and most natural pozzolans the changes in AEA dosage required are generally small, and the dose may even decrease if the materials have a lower fineness than the portland cement. Silica fume significantly increases the amount of AEA required because of its very high fineness, and possibly due to the presence of some unburned carbon in some silica fumes (Carette and Malhotra, 1983). Typically, the amount of AEA required in silica fume concrete is increased by between 125 and 150% of that required in concrete without silica fume.

A rapid screening test, known as the foam index test, was developed by Dodson (Meininger, 1981) to predict when changes in AEA dosages may be required for fly ash concrete. In this test, 16 g of cement is mixed with 4 g of fly ash and 50 ml of water in a glass bottle and shaken vigorously for 1 minute. Small increments of diluted AEA are then added, and the bottle is shaken vigorously for 15 seconds after each addition until a stable foam is

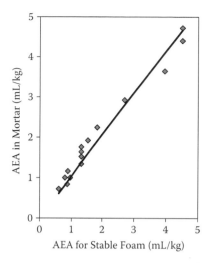

Figure 5.7 Air-entraining admixture requirements (in ml AEA per kg of cement + fly ash) in mortar and foam index test. (Data from Meininger, R.C., *Use of Fly Ash in Air-Entrained Concrete. Report of Recent NSGA-NRMCA Research Laboratory Studies*, National Ready-Mixed Concrete Association, Silver Spring, MD, 1981. Printed with permission from the National Ready-Mixed Concrete Association.)

observed to form on the surface (the foam should remain stable for at least 45 seconds). Figure 5.7 shows the relationship between the amount of AEA required to produce a stable foam in the foam index test and the amount required to produce a certain level of air in mortar mixes (Meininger, 1981).

5.4 SETTING TIME

The impact of SCMs on the setting behavior of concrete is dependent not only on the composition, fineness, and quantity of SCM used, but also on the type and amount of cement, the water-to-cementitious materials ratio (W/CM), the type and amount of chemical admixtures, and the concrete temperature.

It is fairly well established that low-calcium fly ashes extend both the initial and final set of concrete, as shown in Figure 5.8, especially when used at higher levels of replacement and in cold weather. Similarly, high levels of slag can result in slower setting times at low temperatures.

During cold weather, the use of fly ash and slag, especially at high levels of replacement, can lead to very significant delays in both the initial and the final set. These delays may result in placement difficulties, especially with regards to the timing of finishing operations for floor slabs and pavements,

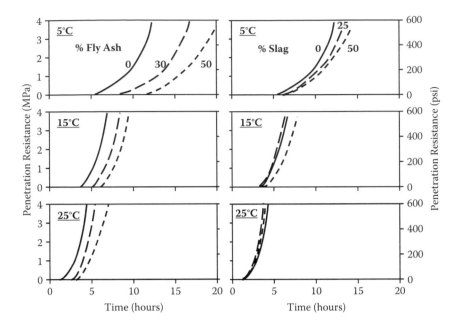

Figure 5.8 Effect of fly ash, slag, and temperature on the penetration resistance of setting concretes proportioned for equal strength at 28 days and workability. (Data from Concrete Society, *The Use of GGBS and PFA in Concrete*, Technical Report 40, Concrete Society, Wexham, Slough, 1991.)

and concrete may require protection to prevent freezing of the plastic concrete. Practical considerations may require that the fly ash or slag content is limited during cold-weather concreting. The use of set-accelerating admixtures may wholly or partially offset the retarding effect of the fly ash or slag. The setting time can also be reduced by using a high early-strength (rapid-hardening) cement or by increasing the initial temperature of the concrete during production in winter (for example, by heating mix water or aggregates).

Higher-calcium fly ashes generally retard setting to a lesser degree than low-calcium fly ashes, probably because the hydraulic reactivity of fly ash increases with increasing calcium content. However, the effect of high-calcium fly ashes is more difficult to predict because the use of some of these ashes with certain cement-admixture combinations can lead to either rapid (or even flash) setting or severely retarded setting (Wang et al., 2006; Roberts and Taylor, 2007).

During hot weather the amount of retardation due to SCMs tends to be small and is likely to be a benefit in many cases.

Silica fume has little impact on the setting behavior of concrete.

The effect of natural pozzolans on setting behavior varies widely depending on, among other factors, the nature and amount of the pozzolan used. For example, highly reactive natural pozzolans such as metakaolin have little impact on the time of setting when used at moderate levels of replacement (10 to 15%), whereas more slowly reactive calcined shales may significantly extend the setting time when used in relatively high amounts (>25%), especially at low temperatures.

With all SCMs testing is required before a new SCM source is introduced to a plant. Testing can determine the effect of the SCM on the setting behavior of concrete produced with the other plant materials. This testing should be conducted at a range of SCM levels and at different temperatures.

Chapter 6

Temperature rise and risk of thermal cracking

6.1 TEMPERATURE RISE

The hydration of portland cement is exothermic and the total heat produced can be as high as 500 kJ/kg. In large concrete pours, where the heat of hydration dissipates very slowly, significant temperature rises in the range of 12 to 14°C per 100 kg (8 to 10°F per 100 lb) cement can occur (FitzGibbon, 1976a, 1976b), which can result in peak internal concrete temperatures as high as 80°C (176°F) or more in certain cases. When concrete cools from its peak temperature it contracts, and thermal cracking can occur if the contraction is restrained externally or if significant thermal gradients occur within the concrete mass (for example, if the external surface cools much more rapidly than the internal bulk). One way to reduce the risk of thermal cracking is to reduce the peak internal temperature of the concrete.

If concrete experiences high internal temperatures during curing the risk of expansion and cracking due to delayed ettringite formation (DEF) will also be increased. The phenomenon of DEF is discussed in Chapter 9 on durability of concrete.

The use of pozzolans that tend to react relatively slowly, such as low-calcium fly ash and certain natural pozzolans, can result in significant reductions in the temperature rise of large concrete pours, and this has long been a major incentive for using such materials in concrete construction. Higher-calcium fly ashes and slag tend to react more quickly than low-calcium fly ash, and are less efficient in reducing temperature rise unless they are used at relatively high levels of replacement. Highly reactive pozzolans such as silica fume and metakaolin have little impact on temperature rise at the levels of replacement normally used.

Much of the published literature on the effects of supplementary cementing materials (SCMs) on the heat generated during the early stages of hydration reports on testing using isothermal calorimetry usually conducted at standard laboratory temperatures. Unfortunately, such testing does not take account of the effect of temperature on the rate of reaction of pozzolans and slag, and since this effect is substantial, such tests often significantly underestimate the heat released by the reaction of these materials at early ages (Bamforth, 1980, 1984).

One of the first full-scale field trials using fly ash was conducted by Ontario Hydro (Mustard and MacInnis, 1959) during the construction of the Otto Holden Dam in Northern Ontario around 1950. Two elements of the dam, measuring $3.7 \times 4.3 \times 11.0$ m ($12 \times 14 \times 36$ ft), were constructed with embedded temperature monitors. One element was constructed using a concrete with 305 kg/m³ (514 lb/yd³) of portland cement, and the other with a concrete with the same cementitious material content but with 30% of the portland cement replaced with a Class F fly ash. Figure 6.1 shows the results from this study, indicating that the use of fly ash reduced the maximum temperature rise over ambient from 47 to 32°C (85 to 58°F).

Bamforth (1980) showed that the temperature rise in foundations was a function of the cementing material content, replacement level of fly ash or slag, and height of the foundation, as shown in Figure 6.2. Although fly ash and slag do reduce the temperature rise, the effects become less pronounced in larger pours (greater lift heights), as the impact of the slower rate of heat development associated with using fly ash or slag becomes less significant in larger pours where the heat dissipates more slowly. However, even in deep foundations the use of fly ash and slag does produce a significant reduction in the peak temperature, as shown in Figure 6.3, which shows the change in temperature with time at the mid-height of three massive foundations 4.5 m (15 ft) deep (Bamforth, 1980). The total concrete volumes placed

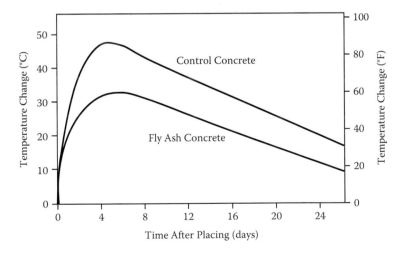

Figure 6.1 Effect of fly ash on temperature rise in concrete dams. (Data from Mustard, J.N., and MacInnis, C., *Engineering Journal*, December 1959, pp. 74–79.)

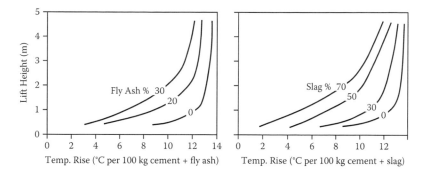

Figure 6.2 Effect of fly ash, slag, and lift height on the temperature rise of massive concrete foundations. (Data from Bamforth, P.B., in *Proceedings of Institution of Civil Engineers*, London, September 1980, Part 2, Vol. 69, pp. 777–800.)

in these foundations were 144, 147, and 212 m³ (188, 192, and 277 yd³), respectively, for the portland cement concrete, fly ash concrete, and slag concrete. The temperature rise above the placing temperature was 54.5°C (from 20.5 to 75°C) for the portland cement concrete. For concrete with 30% low-calcium fly ash the temperature rise was 47.5°C (from 21.5 to 67°C), and for concrete with 75% slag, the temperature rise was 46.0°C (from 18.0 to 64°C).

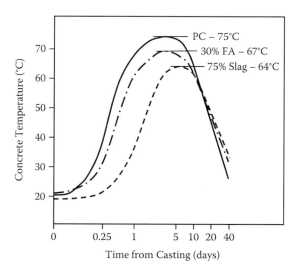

Figure 6.3 Effect of fly ash and slag on temperature at mid-height in large concrete foundations. (Data from Bamforth, P.B., in *Proceedings of Institution of Civil Engineers*, London, September 1980, Part 2, Vol. 69, pp. 777–800.)

Table 6.1 Temperature rise in large concrete blocks produced with HVFA concrete

Mix	Cement kg/m³ (lb/yd³)	Fly ash kg/m³ (lb/yd³)	W/CM	Max. temp. °C (°F)	Time to max. (h)
1	400 (674)	—	0.33	83 (181)	24
2	180 (303)	220 (370)	0.27	54 (129)	96
3	100 (168)	125 (211)	0.49	30 (86)	168

Source: Data from Langley, W.S., et al., ACI Materials Journal, 89(4), 362–368, 1992.

In massive concrete pours where the rate of heat loss is small, the maximum temperature rise in fly ash concrete will primarily be a function of the amount and composition of the portland cement and fly ash used, together with the temperature of the concrete at the time of placing. Concrete with low portland cement contents and high fly ash contents is particularly suitable for minimizing autogenous temperature rises. For example, Langley and coworkers (1992) cast three 3.05 × 3.05 × 3.05 m (10 × 10 × 10 ft) blocks with embedded thermocouples, and showed that the incorporation of 55% fly ash reduced the peak temperature by 29°C (84°F) when the cementitious material content was held constant and by 53°C (127°F) when the total cementitious content was reduced (see Table 6.1). The high-volume fly ash (HVFA) concrete mixes (with ~55% Class F fly ash) were effective in reducing both the rate of heat development and the maximum temperature reached within the concrete block.

Table 6.2 shows data from a later study (Bisaillon et al., 1994) using large monoliths (2.5 × 4.0 × 5.0 m [8.2 × 13.1 × 16.4 ft]) cast with HVFA concrete with Type F fly ash. These results again indicate that the autogenous

Table 6.2 Temperature rise in large concrete monoliths produced with HVFA concrete

Mix	Cement kg/m³ (lb/yd³)	Fly ash kg/m³ (lb/yd³)	W/CM	Strength MPa (psi) 1-day	Strength MPa (psi) 3-day	Max. temp °C (°F)	Time to max. (h)
1	365 (600) Type I	—	0.45	10.3 (1495)	—	68 (154)	29
2	125 (211) Type I	155 (261)	0.46	1.6 (230)	5.1 (740)	44 (111)	53
3	170 (287) Type I	220 (370)	0.29	8.4 (1220)	15.6 (2260)	54 (129)	57
4	330 (556) Type II	—	0.50	7.3 (1060)	14.0 (2030)	55 (131)	75
5	125 (211) Type I	155 (261)	0.41	2.5 (365)	8.4 (1220)	47 (117)	98

Source: Data from Bisaillon, A., et al., ACI Materials Journal, 91(2), 178–187, 1994.

temperature rise can be kept very low with high-volume fly ash when the total cementitious content is kept low (in this case 280 kg/m³ [472 lb/yd³]). This property can be very advantageous when early-age strength is not important. Higher early-age strengths can be achieved by raising the cementitious material content of the HVFA system, although this does result in an increase in the autogenous temperature rise.

HVFA concrete systems have been successfully used in commercial applications to control the temperature rise in large placements (Mehta and Langley, 2000; Mehta, 2002).

Most published work on the effects of fly ash on the rate of heat development and temperature rise in concrete have focused on low-calcium Class F fly ashes. Work by the Bureau of Reclamation (Dunstan, 1984) indicated that the rate of heat development generally increases with the calcium content of the ash. Fly ashes high in calcium may produce little or no decrease in the heat of hydration (compared to plain portland cement) when used at normal replacement levels. Similar results have been reported for studies on insulated mortar specimens (Barrow et al., 1989), where the use of high-calcium ash (>30% CaO) was found to retard the initial rate of heat evolution but did not reduce the maximum temperature rise. However, Carette et al. (1993) reported that there was no consistent trend between ash composition and temperature rise for concretes containing high levels of fly ash (56% by mass of cementitious material). Calcium levels of the ashes used in the study ranged up to 20% CaO. Conduction calorimetry studies conducted at Ontario Hydro in Canada (Thomas et al., 1995) using a wide range of fly ashes (2.6 to 27.1% CaO) showed that the 7-day heat of hydration of cement–fly ash pastes was strongly correlated with the calcium content of the fly ash in agreement with Dunstan (1984). However, these studies also indicated that high-calcium fly ashes could be used to meet performance criteria for low-heat cements (e.g., ASTM C 150 Type IV or ASTM C 1157 Type LH) when used at a sufficient replacement level (Figure 6.4).

High levels of high-calcium (Class C) fly ash have been used to control the temperature rise in mass concrete foundations. One example is the concrete raft foundation for the Windsor Courthouse (Ellis-Don, 1996). This 10,000-m³ (13,000-yd³) concrete raft was 1.2 m (4 ft) thick and was placed in pours 1400 to 1700 m³ (1830 to 2220 yd³) in volume, with placement rates (pumping the concrete) of up to 100 m³/h (130 yd³/h).

Concrete with 50% Class C fly ash was used to control temperature, while thermocouples were used to determine when thermal blankets could be removed without causing thermal shock.

Rivest and coworkers (2004) measured the temperature rise of four concrete monoliths with dimensions 2.5 × 4.0 × 5.0 m (8.2 × 13.1 × 16.4 ft) high. The concretes were produced with (1) 100% Type I cement, (2) 100% Type II (moderate-heat) cement, (3) Type I cement with 56% Class F fly ash (2.0% CaO), and (4) Type I cement with 60% slag. The mixture

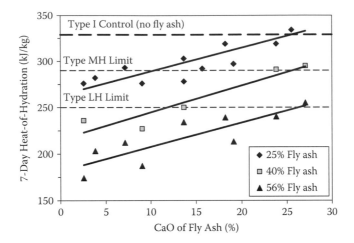

Figure 6.4 Effect of fly ash composition on the heat of hydration using conduction (iso-thermal) calorimetry. (Data from Thomas, M.D.A., et al., in *Proceedings of the Fifth CANMET/ACI International Conference on Fly Ash, Silica Fume, Slag, and Natural Pozzolans in Concrete*, ed. V.M. Malhotra, SP-153, Vol. I, American Concrete Institute, Farmington Hills, MI, 1995, pp. 81–98.)

proportions and temperature data for these blocks are shown in Table 6.3. The Type II cement used was moderate-heat cement that had been specified by Hydro-Quebec for its mass concrete projects. The temperature data confirm previous studies on high-volume fly ash concrete with regards to its ability to lower the temperature rise in massive concrete placements, but the results for slag show that even at a 60% level of replacement there is no significant temperature reduction. The authors (Rivest et al., 2004) suggest that higher temperature rise observed for the slag concrete compared with other published studies is perhaps the result of the very large monoliths used in this study.

Table 6.3 Temperature rise in large concrete monoliths with 56% fly ash or 60% slag

| Mix | Cement kg/m³ (lb/yd³) | SCM kg/m³ (lb/yd³) | W/CM | Temperature °C (°F) | | | Time to max. (h) |
				Initial	Maximum	Rise	
1	365 (608) Type I	—	0.45	22 (72)	68 (154)	46 (83)	29
2	330 (550) Type II	—	0.50	19 (66)	55 (131)	36 (65)	75
3	170 (283) Type I	220 (367) ash	0.32	21 (70)	54 (129)	33 (59)	57
4	132 (220) Type I	190 (317) slag	0.37	24 (75)	69 (156)	45 (81)	71

Source: Data from Rivest, M., et al., in *Proceedings of the 8th CANMET/ACI International Conference on Fly Ash, Silica Fume, Slag, and Natural Pozzolans in Concrete*, ed. V.M. Malhotra, ACI SP-221, American Concrete Institute, Detroit, 2004, pp. 859–878.

Table 6.4 Temperature rise in concrete blocks with various levels of slag

Mix	Slag (%)	Aggregate	Temperature °C (°F)				
			Mean ambient	Maximum center	Maximum surface	Maximum gradient	Time to peak (h)
1	50	Gravel	4.8 (41)	23.5 (74)	15.0 (59)	8.5 (15)	31
2	50	Gravel	4.3 (40)	22.5 (73)	14.0 (57)	8.5 (15)	24
3	50	Limestone	4.7 (40)	24.5 (76)	14.0 (57)	10.5 (19)	34
4	0	Gravel	11.2 (52)	52.4 (126)	34.2 (94)	18.2 (33)	18
5	30	Gravel	10.9 (52)	44.5 (112)	29.1 (84)	15.4 (28)	24
6	70	Gravel	11.1 (52)	24.7 (76)	18.1 (65)	6.6 (12)	42
7	50	Limestone	6.8 (44)	26.3 (79)	17.5 (64)	8.8 (16)	28
8	70	Limestone	7.0 (45)	20.4 (69)	16.8 (62)	3.6 (6)	48
9	50	Limestone	6.9 (44)	28.6 (83)	22.3 (72)	6.3 (11)	40

Source: Data from Osborne, G.J., and Connell, M.D., in *Proceedings of the 5th CANMET/ACI International Conference on the Durability of Concrete*, ACI SP-192, Vol. 1, American Concrete Institute, Detroit, 2000, pp. 119–139.

Osborne and Connell (2000) demonstrated that slag is considerably more effective in reducing the temperature rise in smaller specimens. In their study, 750-mm (30-in.) concrete cubes were cast in 19-mm (3/4-in.) -thick plywood forms insulated with 50-mm (2-in.) -thick polystyrene, and a 10-mm-thick "ethafoam" matting plus 50 mm (2 in.) of polystyrene was placed on the top surface. Temperature data are presented in Table 6.4. All mixes contained approximately 395 kg/m³ (658 lb/yd³) of cementing material consisting of various blends of ordinary portland cement and slag. Five mixes were cast using a gravel aggregate (including the control without slag), and the remaining four with a crushed limestone coarse aggregate.

6.2 RISK OF THERMAL CRACKING

The risk of thermal cracking depends on the magnitude and rate of temperature changes, the magnitude of temperature gradients within the concrete element, the thermal properties of the concrete (particularly the coefficient of thermal expansion), and the mechanical properties of the concrete (tensile strength and strain capacity, stiffness, and creep). When concrete cools it will contract; however, if the movement of the concrete is restrained, tensile stresses will build up inside the concrete element. If the concrete is completely restrained from movement, the stresses can be calculated using elastic analysis as follows:

$$\sigma_t = E \cdot \alpha \cdot \Delta T \tag{6.1}$$

where σ_t = thermal (tensile) stress, E = modulus of elasticity, α = coefficient of thermal expansion, and ΔT = temperature change.

This assumes that any compressive stresses that develop when the concrete is increasing in temperature are negligible. More realistically, the thermal stress should be calculated with allowances being made for the degree of restraint and the effects of stress relief due to creep as follows:

$$\sigma_t = K_r \cdot \frac{E}{1+\phi} \cdot \alpha \cdot \Delta T \qquad (6.2)$$

where K_r = restraint factor (K_r = 0 to 1) and φ = creep coefficient.

If a concrete is completely free to move, there is no restraint, K_r = 0, and thermal stresses do not develop. In practice, the concrete will be restrained either externally due its bond with the foundation or adjacent concrete members or internally due to differential cooling rates (i.e., the surface cools more rapidly than the core of a massive member). Values of K_r for various conditions are given in ACI 207.2R (ACI 207, 1995).

Concrete will crack when the following condition is met:

$$K_r \cdot \frac{E}{1+\phi} \cdot \alpha \cdot \Delta T > f_t' \qquad (6.3)$$

where f_t' = tensile strength of the concrete.

The use of SCMs in concrete has little or no effect on K_r and α, but will influence E, φ, ΔT, and f_t'. It has been shown that the judicious use of SCMs, particularly Class F fly ash and other slowly reacting pozzolans, can reduce the temperature rise, and hence ΔT. But the use of these materials will also reduce the strength of concrete at early ages, and thus its ability to resist tensile stresses. However, the reduced strength is likely offset by a reduced stiffness and increased creep at early ages, which, in addition to the reduced temperature, will serve to reduce the magnitude of the thermal stresses that develop on cooling. A recent study confirmed that concrete containing fly ash, especially low-lime fly ash, showed a lower risk of early-age cracking primarily due to the lower temperature rises and increased early-age creep (Riding et al., 2008).

Chapter 7

Mechanical properties

7.1 COMPRESSIVE STRENGTH

The compressive strength of concrete containing supplementary cementing materials (SCMs) depends on a large number of factors, including the type and amount of the SCM used, the composition and amount of the portland cement, other mixture proportions (especially W/CM and air content), age, temperature, and curing.

Figure 7.1 shows typical strength–time relationships for concrete cylinders with W/CM = 0.45 cured in lime-saturated water at normal laboratory temperature. The concretes were produced with various SCMs using replacement levels typical for that particular SCM. In these concretes, the cement was replaced with an equal mass of SCM, the sand content was then adjusted to maintain unit volume, and the dosages of air-entraining and water-reducing admixtures were modified to maintain the desired air content (5 to 8%) and slump (75 to 125 mm, 3 to 5 in.). No other changes to the mixture proportions were made. Under these conditions, the use of 25% Class F fly ash reduces the early-age strength of the concrete, but increases the long-term strength. The fly ash concrete achieves parity with the control concrete without SCM sometime between 28 and 91 days. Concrete with Class C fly ash behaves in a similar manner, but the extent of the early-age strength reduction is less and strength parity is achieved at an early age (typically 14 to 28 days). The use of 35% slag also reduces early-age strength, but to a lesser extent than fly ash, and equal strength is often seen as early as 7 days. The later-age strength of slag concrete is also improved compared to the control. Concrete with silica fume behaves in a different manner because of the relatively rapid reaction of silica fume, which is attributed to its very fine particle size. Strengths at 3 days and beyond are usually improved significantly by the presence of silica fume. The behavior of concrete containing natural pozzolans will depend on the nature of the pozzolan. Concrete containing relatively slowly reactive pozzolans, such as some calcined shales and volcanic ashes, will behave in a similar manner as concrete with Class F fly ash, whereas concrete with fast-reacting pozzolans

77

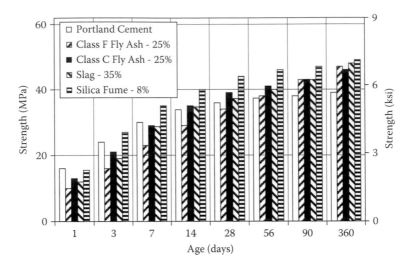

Figure 7.1 Strength development for air-entrained concretes with typical levels of SCMs (W/CM = 0.45).

such as metakaolin or rice hush ash will show strength development similar to that of concrete containing silica fume.

Concrete is often proportioned to achieve a certain 28-day compressive strength (as determined on standard-cured cylinders) to meet the requirements of a specification. In such cases, it may be necessary to proportion concretes containing SCMs to have different W/CM values. It has long been known that for any given set of materials, the strength of the concrete is largely dependent on W/CM (Abrams, 1918). Figure 7.2 illustrates how the relationship between W/CM and 28-day strength might vary for mixtures with different cementing materials. At a given W/CM, concrete produced with silica fume has the highest strength and concrete with fly ash the lowest strength. To achieve equal 28-day strength, the W/CM for concrete with fly ash has to be reduced compared with the concrete without SCM, whereas that for the concrete with silica fume can be increased. If higher amounts of fly ash were to be used, the W/CM would need to be reduced even further. Concrete with moderate levels of slag (25 to 40%) would not generally need an adjustment to be made to the W/CM to achieve a similar 28-day strength as concrete without SCM. However, at higher replacement levels (≥50%) the W/CM would have to be reduced.

Reductions in the W/CM can, of course, be achieved by increasing the cementitious material content, decreasing the water content, or both. As good quality fly ash reduces the water demand of concrete, a certain reduction in the water content can be made without losing slump. The addition of water-reducing admixtures will allow further reductions in the water

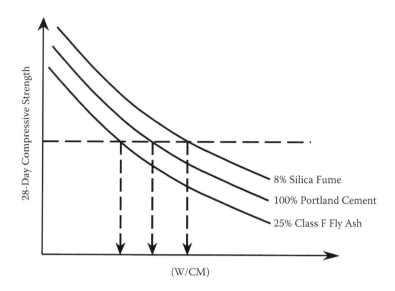

Figure 7.2 Relationship between W/CM and strength for different concrete mixes.

content, and it is often possible to achieve the desired W/CM without increasing the cementitious material content above that used in a mixture of the same strength but without SCM.

Another method for compensating for the relatively slow early-age strength development of concrete with slowly reacting pozzolans is to use them in conjunction with more rapidly reacting pozzolans such as silica fume. Figure 7.3 illustrates how modifying the W/CM and adding silica fume can compensate the low early-age strength of concrete containing high levels of Class F fly ash.

Ternary cements containing silica fume and slag can also be used to produce concrete with early-age strength that is comparable to concrete produced without SCM. Figure 7.4 shows the effect of W/CM on the strength of concrete produced with a factory blend of 4% silica fume and 22% slag compared with similar concrete produced with portland cement as the only binder (Thomas et al., 2007). The cements were produced using the same portland cement clinker. The early-age strengths are similar, but the 28-day strengths are increased in the concrete produced with the ternary cement.

SCMs have been widely used in high-strength concrete for high-rise construction, especially silica fume, or blends of silica fume with either fly ash or slag. Table 7.1 shows some typical examples where blends of SCMs have been used in high-strength concrete structures. Although silica fume has become synonymous with high-strength concrete, it is possible to produce to high strengths (≥70 MPa or 10,000 psi) without silica fume when it is not available. Aitcin and Neville (1993) cite an example where concrete with a

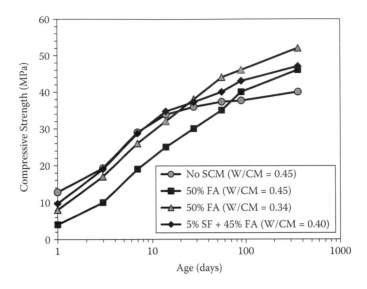

Figure 7.3 Adjusting W/CM and adding silica fume to improve the early-age strength of concrete with high levels of Class F fly ash.

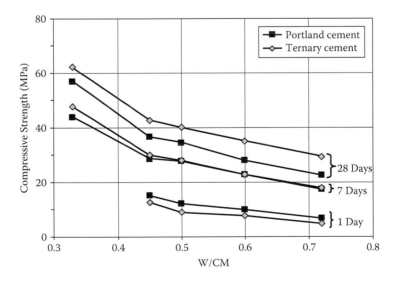

Figure 7.4 Effect of W/CM on strength development of concrete produced with a ternary cement containing silica fume and slag.

Table 7.1 Examples of high-strength concrete applications using SCMs

Project	Ref.	Date	SCM	W/CM	Strength, MPa (psi) at age
Water Tower Place	1	1970	15% FA	—	60 (8,700) specified
Scotia Plaza	2	1987	7.5% SF 28% slag	0.30	61.8 (8,960) at 2 days 83.7 (12,140) at 28 day 93.6 (13,580) at 91 days
Two Union Square	3	1988	7.7% SF	0.25	119 (17,200) at 28 days 145 (21,000) at 91 days
Petronas Towers	4		6% SF 13% FA	0.27	97 (14,070) at 28 days
Key Corp Tower	4		7.5% SF 27% slag	0.24	87 (12,620) at 7 days 98 (14,210) at 28 days 104 (15,080) at 56 days
Trump Tower	5		Various amounts of SF, FA, and slag	0.26–0.28	55.2 (8,000) at 7 days 65.5 (9,500) at 28 days 110 (16,000) at 90 days

Note: SF = silica fume, FA = fly ash.

References: 1 = Aitcin (1998), 2 = Ryell and Bickley (1987), 3 = Anon, 4 = ACI 234 (2006), 5 = M. Pistilli (unpublished).

strength of 90 MPa (13,050 psi) at 28 days and 111 MPa (16,100 psi) at 91 days was produced in a ready-mix plant with 26% fly ash as the only SCM. Furthermore, the Joigny Bridge in France was built using concrete without any SCM, which had a strength of 79 MPa (11,460 psi) at 91 days (Aitcin and Neville, 1993).

7.1.1 Effect of temperature

The rate of reaction of pozzolans and slag is strongly influenced by temperature, and consequently, curing temperature has a significant impact on the strength development of concrete produced with these materials. Figure 7.5 shows the strength development to 28 days of a conventional concrete without SCM and a concrete produced with 50% slag when cured at either 5, 20, or 30°C (41, 68, or 86°F). At 20°C (68°F), the slag concrete gains strength at a slightly slower rate than the concrete without slag at ages up to 7 days, but achieves a similar strength by 28 days. At 30°C (86°F) differences between the two concretes are less marked at early ages and the slag achieves a higher strength after 28 days. At 5°C (41°F), the strength gain of the slag concrete is retarded and the strength of the slag concrete is significantly less than the control even after 28 days.

Concrete containing fly ash responds to temperature in a manner similar to that of concrete containing slag. Figure 7.6 shows the effect of curing temperature (above 20°C, 68°F) on the 28-day strength of concrete with and without fly ash. Curing concrete without SCM at elevated temperature

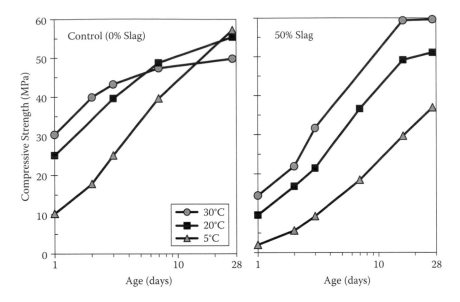

Figure 7.5 Effect of curing temperature on the strength development of slag concrete. (Data from Wainwright, P.J., in *Cement replacement materials*, ed. R.N. Swamy, Concrete Technology Design Vol. 3, Surrey University Press, Guildford, Surrey, UK, 1986, pp. 100–133.)

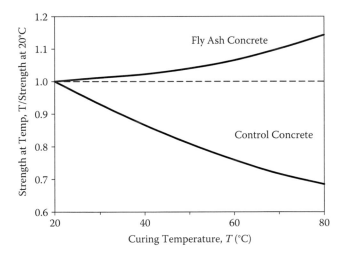

Figure 7.6 Effect of temperature rise (above 20°C, 68°F) on the 28-day strength of concrete. (Data from Williams, J.T., and Owens, P.L., in *Proceedings of the International Symposium on the Use of PFA in Concrete*, ed. J.G. Cabrera and A.R. Cusens, Department of Civil Engineering, University of Leeds, Leeds, UK, 1982, pp. 301–313.)

increases the early-age strength (e.g., at 1 to 3 days) but impairs the long-term strength. The data in Figure 7.6 suggest that the 28-day strength of portland cement concrete cured at 80°C (176°F) may be some 30% lower than the 28-day strength of the same concrete cured at 20°C (68°F). This is due to the poor distribution of the rapidly forming hydration products at elevated temperature, which actually leads to a more porous microstructure. In concrete containing fly ash (and other SCMs) these effects are offset by the increased rate of reaction of the SCM leading to the production of more hydrates, and the higher-temperature curing actually leads to increases in the 28-day strength.

7.1.2 Temperature-matched curing

Given the significant effect that temperature has on the rate of strength gain, it is clear that *in situ* strength development (of concrete in a structure) is likely very different from the strength development of standard-cured test specimens produced from the same mix. In thick sections, temperatures may be in excess of 50°C (120°F), and the rate of strength gain will be accelerated compared with test specimens, whereas in thin sections cast in cold weather, the rate of strength gain will be retarded.

Figure 7.7 shows strength data from a study (Bamforth, 1980) where 100-mm cubes were subjected either to standard laboratory curing conditions or to temperature-matched curing. The cubes were produced during the placement of three massive foundations using mixes either without

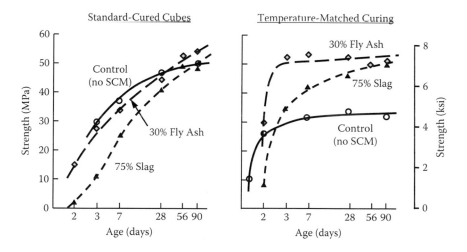

Figure 7.7 Strength development of 100-mm cubes subjected to standard laboratory curing and temperature-matched curing. (Data from Bamforth, P.B., in *Proceedings of Institution of Civil Engineers*, London, September, Part 2, Vol. 69, 1980, pp. 777–800.)

SCM, with 30% fly ash, or with 75% slag. The temperature-matched cubes were cured following the same temperature cycle as that recorded at the midpoint of the foundation pour; the temperature-time data for the foundations are presented in Figure 6.3. The results for the standard-cured cubes show similar strengths for all mixes at 28 days (the mixes were proportioned to achieve the same strength at 28 days), with the concretes containing SCMs having lower strengths at earlier ages and higher strengths at later ages. The results for the temperature-matched curing show that the concrete without SCM showed significant strength loss due to the elevated temperature, with the fly ash and slag concrete having improved strength compared to the control concrete at all ages beyond 3 days.

7.1.3 Cementing efficiency (or *k* value) concept

The concept of a cementing efficiency factor, k, was first proposed by Smith (1967) as the basis of his rational proportioning approach for fly ash concrete. The aim of the approach was to permit fly ash concretes to be proportioned using Abrams' relationship between strength and water-cement ratio. It was understood at this time that if fly ash were to be used as a partial replacement for portland cement without adversely affecting the 28-day compressive strength, the mass of fly ash added to the concrete would have had to be in excess of the mass of cement removed. Adjustments to the aggregate (usually sand) and water content would then be made to keep the volume and workability constant. Originally, the k factor was defined as the mass of portland cement that makes the same contribution to the strength of concrete as a unit mass of fly ash.

Using this concept, concretes will have similar 28-day strengths, provided they are produced with the same water-to-effective cement ratio, which is defined as

$$\text{Effective W/CM} = \frac{W}{c + k \cdot p} \tag{7.1}$$

where
  W = water content
  c = portland cement content
  p = pozzolan content
  k = cement efficiency factor for the pozzolan

Figure 7.8 shows the relationship between the 28-day compressive strength (100 mm, 4 in. cube specimens) of concrete containing fly ash plotted against W/CM and against the effective W/CM using $k = 0.32$

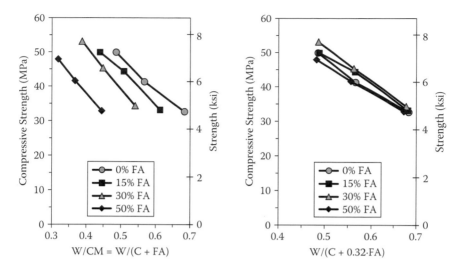

Figure 7.8 Relationship between 28-day strength and water-to-effective cement ratio for concrete containing fly ash. (Data from Thomas, M.D.A., and Matthews, J.D., *Durability of PFA Concrete*, BRE Report BR216, Building Research Establishment (Department of the Environment), Watford, UK, 1994.)

(Thomas and Matthews, 1994). According to the *k* concept, 1 kg of fly ash makes the same contribution to the 28-day strength as 0.32 kg of portland cement. Studies on silica fume, on the other hand, have shown that the value for *k* typically ranges from 2 to 4, the value increasing with the total cementitious material content, and decreasing with the level of silica fume (Sellevold and Nilsen, 1987). Little data are available on the efficiency factor for slag, but general experience indicates that the *k* value of unity is probably appropriate.

In the European Standard for concrete (EN 206-1:2000) *k* values are used for fly ash and silica fume. For fly ash the *k* value varies depending on the strength class of the cement as follows:

For CEM 1 32.5 cement: *k* = 0.2
For CEM 1 42.5 and higher: *k* = 0.4

The maximum amount of fly ash to be included in the *k* value concept is

Fly ash/Cement ≤ 0.33 by mass

This translates to a maximum of 25% fly ash when expressed as a percentage of the total cementitious content (cement plus fly ash). If a greater amount of fly ash is present in the concrete, the amount above this limit cannot be used in the calculation of the effective W/CM (in accordance with Equation 7.1) or the minimum cement content.

For silica fume, the k value in the European Standard depends on the water/cement ratio as follows:

For water/cement ≤ 0.45: $k = 2.0$
For water/cement > 0.45: $k = 2.0$, except for exposure classes XC and XF, then $k = 1.0$

Exposure classes XC and XF refer to cases where concrete is exposed either to conditions that may lead to carbonation-induced corrosion of the steel (XC) or to freeze-thaw attack (XF). The amount of silica fume that can be used in calculating the effective W/CM and the minimum cement content is limited as follows:

Silica fume/Cement ≤ 0.11 by mass

This translates to a maximum of 10% silica fume when expressed as a percentage of the total cementitious content (cement plus silica fume).

No k value is given for slag or natural pozzolans in the European Standard.

The author's own research on fly ash (Thomas and Matthews, 1994) does not support the k value concept, as the k value or efficiency of fly ash varies depending on (1) the composition of the fly ash, (2) the level of replacement used, (3) the composition of the portland cement, and (4) the age at which the strength is determined. Furthermore, fly ash or any other SCM may be more efficient in terms of its contribution to one property than another. For example, SCMs make a significant contribution to reducing the permeability of concrete (as discussed in Chapter 9), and for concretes of equal strength, those containing SCM can be expected to have a lower permeability than concrete without SCM. Consequently, the use of the k concept is not recommended for the design or specification of concrete mixtures.

7.2 FLEXURAL AND TENSILE STRENGTH

There are various equations for predicting the flexural and tensile strength of concrete from the compressive strength. Such relationships include the following:

	SI Units	Inch-Pound Units	
ACI 318	$f_r' = 0.6\sqrt{f_c'}$	$f_r' = 7.5\sqrt{f_c'}$	Equation 7.2a and b
Mindess et al., 2002	$f_r' = 0.438 f_c'^{2/3}$	$f_r' = 2.30 f_c'^{2/3}$	Equation 7.3a and b
	$f_{sp}' = 0.305 f_c'^{0.55}$	$f_{sp}' = 4.34 f_c'^{0.55}$	Equation 7.4a and b

where

f_c' = compressive strength (MPa or psi)

f_r' = flexural strength or modulus of rupture (MPa or psi)

f_{sp}' = splitting strength (MPa or psi)

These empirical equations have been developed from data for portland cement contents, but they are generally considered to be appropriate for concrete containing SCMs.

7.3 MODULUS OF ELASTICITY

The modulus of elasticity, E_c, can also be predicted from the compressive strength, f_c', of concrete, and the equations used in ACI 318 are as follows:

	SI Units	Inch-Pound Units	
Normal weight concrete	$E_c = 4,730\sqrt{f_c'}$	$E_c = 57,000\sqrt{f_c'}$	Equation 7.5a and b
Lightweight concrete	$E_c = w_c^{1.5} 0.043\sqrt{f_c'}$	$E_c = w_c^{1.5} 33\sqrt{f_c'}$	Equation 7.6a and b

where

w_c = unit weight of concrete (kg/m³ or lb/ft³)

E_c = modulus of elasticity (MPa or psi)

The relationship between the modulus of elasticity and the compressive strength of concrete is essentially unaffected by the presence of SCMs, and the predictive equations are equally applicable to concrete with or without SCM. Figure 7.9 shows strength and modulus data for 21 concrete mixtures with fly ash levels up to 50% (Ghosh and Timusk, 1981), and these data indicate that fly ash concrete can be expected to have a similar modulus as portland cement concrete of equivalent strength. The modulus determined by testing in this study was consistently higher than the modulus predicted using the equations in ACI 318.

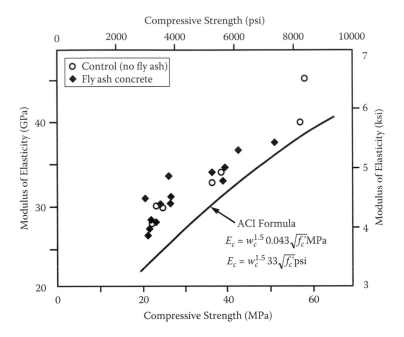

Figure 7.9 Relationship between strength and modulus of elasticity for concrete with and without fly ash. (From Ghosh, R.S., and Timusk, J., *ACI Materials Journal*, 78(5), 351–357, 1981. Printed with permission from the American Concrete Institute.)

The use of moderate to high levels of fly ash and slag may be expected to reduce the modulus of elasticity at early ages, but increase the modulus at later ages. This effect is shown in Figure 7.10, which has been plotted from data reported by Wainwright and Tolloczko (1986).

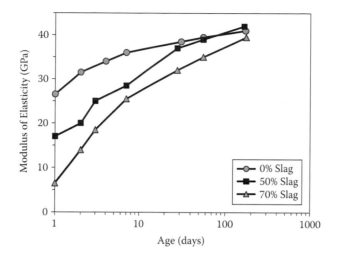

Figure 7.10 Effect of slag on the modulus of elasticity of concrete. (Data from Wainwright, P.J., and Tolloczko, J.J.A., in *Proceedings of the Second CANMET/ACI International Conference on Fly Ash, Silica Fume, Slag, and Natural Pozzolans in Concrete*, ed. V.M. Malhotra, ACI SP-91, Vol. 2, American Concrete Institute, Farmington Hills, MI, 1986, pp. 1293–1322.)

Chapter 8

Volume stability

This chapter discusses the influence of supplementary cementing materials (SCMs) on volume changes due to chemical and autogenous shrinkage, drying shrinkage, temperature changes, and sustained load (creep). Elastic deformations were discussed in the previous chapter, and volume changes resulting from various deterioration mechanisms (e.g., alkali-aggregate reaction and sulfate attack) will be discussed in the next chapter.

8.1 CHEMICAL AND AUTOGENOUS SHRINKAGE

When cement hydrates there is an increase in the volume of solids, as the volume of the hydration products is greater than the volume of the unhydrated cement. This leads to a reduction in the porosity of the paste, resulting in an increase in strength and reduction in permeability. However, the volume of the hydration products is actually less than the volume originally occupied by the unhydrated cement plus the water consumed by the hydration process. This means that the absolute volume of solids plus water is reduced by the hydration process. The reduction in this volume is referred to as chemical shrinkage. In a sealed system chemical shrinkage will lead to some reduction in the overall volume of the paste, as shown in Figure 8.1 (Kosmatka and Wilson, 2011). Prior to initial set, the reduction in volume of the cement plus water is manifested as a reduction in the volume of the paste. As the cement paste sets and begins to gain strength, there is an increasing restraint to further volume change, and any further chemical shrinkage results in a combination of the removal of water from some of the voids within the paste and some reduction in the total volume of the paste. The actual reduction in the volume of the paste is referred to as autogenous shrinkage. The autogenous shrinkage is less than the chemical shrinkage because of the restraint to volume change after setting. The removal of water from some of the voids in the paste results in a reduction in the internal relative humidity, and this is referred to as self-desiccation.

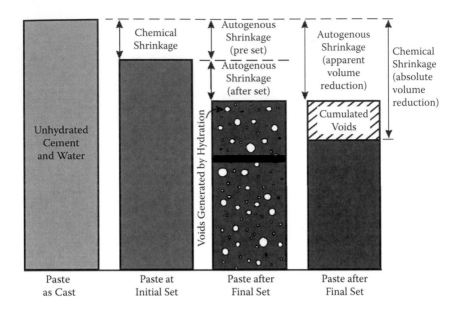

Figure 8.1 Chemical shrinkage and autogenous shrinkage of cement paste—not to scale. (From Kosmatka, S.H., and Wilson, M.L., *Design and Control of Concrete Mixtures*, EB001, 15th ed., Portland Cement Association, Skokie, IL, 2011. Reproduced with permission of the Portland Cement Association.)

Chemical shrinkage cannot be prevented; it occurs as a direct result of the chemical reactions that occur between cement and water. However, self-desiccation and autogenous shrinkage can be minimized by providing an external source of water to the paste (or concrete) as soon as the hydration starts (Aitcin, 1999a, 1999b). Internal curing using water-saturated lightweight aggregate or water-absorbent polymers can further reduce the potential for self-desiccation and autogenous shrinkage.

Autogenous shrinkage in conventional concrete is considered to be negligible (<100 μs) compared to drying shrinkage; however, it increases as the W/CM decreases (Figure 8.2) and can become a significant component of the overall shrinkage in concrete with W/CM < 0.42. Experimental studies on concrete with W/CM = 0.30 have shown that the autogenous shrinkage can be in the range of 200 to 400 μs, which is comparable to drying shrinkage strains (Tazawa and Miyazawa, 1995). Autogenous shrinkage of concrete also decreases as the volume of paste in the concrete mixture decreases (and aggregate content increases).

Data from studies on autogenous shrinkage should be interpreted with caution. Many studies measure the volume change of sealed specimens starting at 1 day. However, significant autogenous shrinkage can occur during the first 24 hours, especially for concrete with low W/CM (Aitcin,

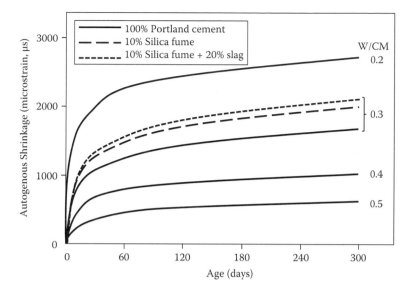

Figure 8.2 Effect of W/CM, silica fume, and slag on the autogenous shrinkage of cement paste. (Data from Jiang, Z., et al., *Cement and Concrete Research*, 35, 1539–1545, 2005. Printed with permission from the American Concrete Institute.)

1999a, 1999b). Also, tests conducted on pastes will result in autogenous shrinkage values many times (about one order of magnitude) greater than tests conducted on concretes.

The use of silica fume has been shown to increase the autogenous shrinkage of concrete compared with concrete without SCM produced at the same W/CM. Figure 8.2 shows the autogenous shrinkage of pastes and the effect of W/CM and 10% silica fume (Jiang et al., 2005). Figure 8.3 shows a similar effect of silica fume on the autogenous shrinkage of concrete (Zhang et al., 2003). Figure 8.4 shows data for slag concretes that indicate that the autogenous shrinkage increases with increasing slag contents. There are few data available on the effects of fly ash and other pozzolans on concrete.

8.2 DRYING SHRINKAGE

Drying shrinkage refers to the reduction in volume caused by the loss of water from hardened concrete due to evaporation. The strains produced by drying are significant (typically >400 μs) and must be accounted for in design and construction. For example, contraction joints have to be provided in pavements, driveways, and slabs to prevent uncontrolled cracking of the concrete when it shrinks. Uneven drying can also lead to warping and curling of concrete slabs. The shrinkage of concrete will also lead to

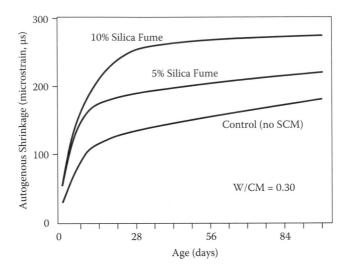

Figure 8.3 Effect of silica fume on the autogenous shrinkage of concrete. (From Zhang, M.H., et al., *Cement and Concrete Research*, 33(10), 1687–1694, 2003. Printed with permission by Elsevier.)

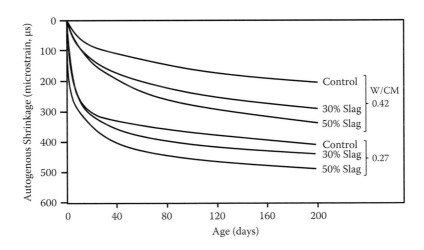

Figure 8.4 Effect of slag on the autogenous shrinkage of concrete. (From Lee, K.M., et al., *Cement and Concrete Research*, 36(7), 1279–1285, 2006. Printed with permission from Elsevier.)

reductions in prestressing, and these losses have to be accounted for in the design stage.

There are many factors that affect cracking, the principal ones being water content, aggregate volume, aggregate type, water-cement ratio, moist curing period, and surface area-to-volume ratio. To minimize the drying shrinkage in concrete, the following strategies should be considered:

- Maximize the aggregate content (increase maximum aggregate size, optimize grading).
- Reduce the unit water content (judicious use of water-reducing admixtures, fly ash).
- Reduce the W/CM.
- Extend the period of moist curing.
- Use shrinkage-reducing admixtures.

The water content and aggregate volume are perhaps the most important parameters with regards to mixture proportioning, and there is a significant increase in the drying shrinkage as either the water content is increased or the aggregate volume decreased.

The impact of supplementary cementing materials on drying shrinkage is secondary compared to the factors listed above. In general, the drying shrinkage of properly proportioned and adequately cured concrete containing SCM will be similar or slightly lower than that of equivalent concrete without SCM (Figure 8.5), although there is some evidence that slag may increase shrinkage slightly.

8.3 CREEP

Creep is the term used to describe time-dependent deformation under a sustained load (Figure 8.6). The amount of creep exhibited by a concrete mixture will depend on the composition of the mixture and the nature of the loading. The water content, W/CM, and aggregate volume of the concrete have the biggest influence on creep; the type of cement has a secondary effect. To minimize the creep of concrete, the following adjustments to mixture proportioning should be considered:

- Maximize the aggregate content (increase maximum aggregate size, optimize grading).
- Reduce the unit water content (judicious use of water-reducing admixtures, fly ash).
- Reduce the W/CM.
- Increase maturity prior to loading (delay loading, steam curing).

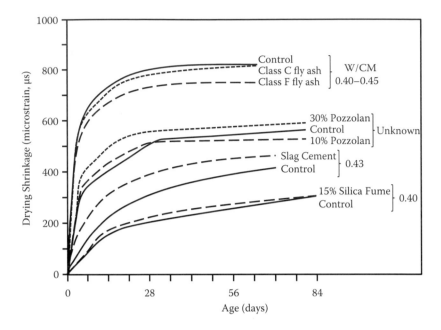

Figure 8.5 Effect of SCMs on drying shrinkage of concrete. Data for fly ash concretes
from Gebler and Klieger (1986); the graphs represent the average of four
different Class C fly ashes and six different Class F fly ashes at a replacement
level of 25%. Data for concrete with pozzolan (volcanic ash) from Mehta
(1981). Data for slag cement from Brooks et al. (1992); the graph for slag
cement is the average for mixes with 30, 50, and 70% slag. Data for silica
fume from Malhotra et al. (1987). (Data from Gebler, S.H., and Klieger, P.,
Effect of Fly Ash on Some of the Physical Properties of Concrete, Research and
Development Bulletin RD089, Portland Cement Association, Skokie, IL, 1986;
Mehta, P.K., *Cement and Concrete Research*, 11, 507–518, 1981; Brooks, J.J., et
al., in *Proceedings of the Fourth International Conference on Fly Ash, Silica Fume,
Slag, and Natural Pozzolans in Concrete*, SP-132, Vol. 2, American Concrete
Institute, Farmington Hills, MI, 1992, pp. 1325–1341; Malhotra, V.M., et al.,
Condensed Silica Fume in Concrete, CRC Press, Boca Raton, FL, 1987.)

For a given concrete mixture the amount of creep will depend on the
loading stress-to-concrete strength ratio (stress-strength ratio), the duration of the load, and whether the concrete is sealed or permitted to dry
during loading.

The creep coefficient, C, of concrete is defined as the ratio of the creep
strain, ε_{cr}, to the elastic strain, ε_e, as follows:

$$C = \frac{\varepsilon_{cr}}{\varepsilon_e} \tag{8.1}$$

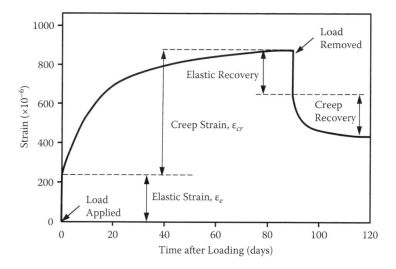

Figure 8.6 Time-dependent deformation (creep) of concrete. (From Mindess, S., et al., *Concrete*, 2nd ed., Prentice Hall, Englewood Cliffs, NJ, 2003. Printed with permission from Pearson Ed.)

Specific creep, Ø, is the ratio of the creep strain, ε_{cr}, to the applied stress, σ, as follows:

$$\varnothing = \frac{\varepsilon_{cr}}{\sigma} \tag{8.2}$$

Typical values for specific creep are 150 µs/MPa or 1 µs/psi (Mindess et al., 2003).

The effect of testing creep under sealed or drying conditions is shown in Figure 8.7. If a concrete is tested under simultaneous loading and drying conditions, the total deformation that occurs is greater than the sum of the drying shrinkage, ε_{sh} (determined under the same drying conditions but with no load), and the basic creep, ε_{bc} (determined under the same load, but with the concrete in a sealed condition). The excess deformation is termed drying creep, ε_{dc}.

There are conflicting data in the literature concerning the effects of pozzolans and slag on the creep behavior of concrete, and much of this stems from differences in the testing regimes used. For example, if loaded at early age, concrete containing moderate to high levels of fly ash or slag may be expected to creep more than portland cement concrete under the same load because of the lower strength of the fly ash and slag concrete at early age (creep increases with stress-strength ratio). However, at later ages the creep will be less in the fly ash and slag concrete because of their higher strength

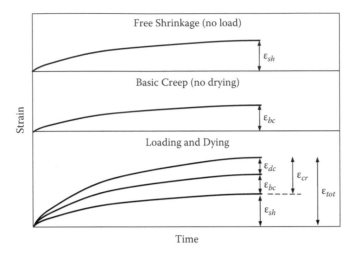

Figure 8.7 Creep of concrete under sealed and drying conditions. (From Mindess, S., et al., *Concrete*, 2nd ed., Prentice Hall, Englewood Cliffs, NJ, 2003. Printed with permission from Pearson Ed.)

at later ages (lower stress-strength ratio). An example of this effect is shown in Figure 8.8 for slag concrete (Brooks et al., 1992). In this study, concrete was produced with slag from four different sources (United Kingdom, Belgium, South Africa, and Japan), and sealed concrete specimens were loaded at a stress-strength ratio of 0.2 at 14 days.

Figure 8.9 shows basic creep (sealed specimens) and total creep (simultaneous loading and drying) for concretes with silica fume and slag (Li et al., 2002). Concrete mixtures were produced with W/CM = 0.30 and binders consisting of 100% portland cement, 10% silica fume, 65% slag, and a blend of 10% silica fume and 55% slag. A second control was produced with W/CM = 0.50. The specimens were loaded to 30% of the strength at 3 days and stored at 30°C. The specific creep for the sealed specimens is significantly reduced by the inclusion of 10% silica fume and by the combination of 10% silica fume plus 55% slag. The concrete with 65% slag showed higher creep at early ages, but the specific creep at 120 days was the same as the control with W/CM = 0.30. The specific creep under drying conditions was reduced for all mixes with SCM compared with the controls.

Figure 8.10 shows the results of creep tests for 21 concrete mixtures with fly ash levels up to 50% (Ghosh and Timusk, 1981); the concretes were proportioned to have 28-day strengths of either 20, 35, or 55 MPa (3000, 5000, or 8000 psi). Specimens were loaded to 30% of the compressive strength at 28 days and were subjected to simultaneous loading and drying. For a given stress-strength ratio fly ash concrete shows lower total

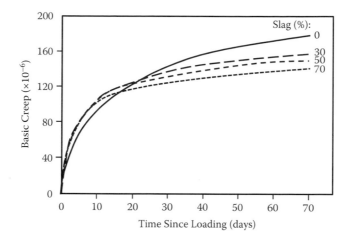

Figure 8.8 Effect of slag on basic creep of sealed concrete specimens. (From Brooks, J.J., et al., in *Proceedings of the Fourth International Conference on Fly Ash, Silica Fume, Slag, and Natural Pozzolans in Concrete*, SP-132, Vol. 2, American Concrete Institute, Farmington Hills, MI, 1992, pp. 1325–1341. Printed with permission from the American Concrete Institute.)

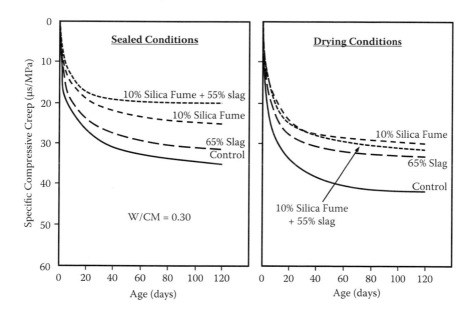

Figure 8.9 Effect of silica fume and slag on specific creep measured under sealed and drying conditions. (From Li, H., et al., *ACI Materials Journal*, 99(1), 3–10, 2002. Printed with permission from the American Concrete Institute.)

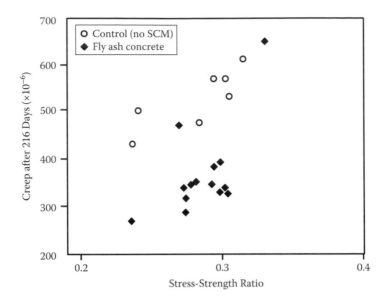

Figure 8.10 Effect of fly ash on creep. (From Ghosh, R.S., and Timusk, J., *ACI Materials Journal*, 78(5), 351–357, 1981. Printed with permission from the American Concrete Society.)

creep strain after 216 days. Ghosh and Timusk (1981) attributed this to the significant strength gain exhibited by the fly ash after loading. Also, for concrete of the same target 28-day strength, the fly ash concretes had lower W/CM.

Based on tests on concretes with fly ash or silica fume, Luther and Hansen (1989) proposed that there is no significant difference between the specific creep of portland cement concrete, silica fume concrete, and fly ash concrete of the same strength. Indeed, this is consistent with the statement in ACI 232.2R that "the effects of fly ash on creep strain of concrete are limited primarily to the extent to which fly ash influences the ultimate strength." This statement can likely be extended to all SCMs.

8.4 TEMPERATURE CHANGES

Concrete will expand and contract as the temperature rises and falls. A typical value for the coefficient of thermal expansion of concrete is 10 μs/°C (5.5 μs/°F). The actual value may range from 6 to 13 μs/°C (3.2 to 7.0 μs/°F), depending on the type and content of aggregate, cement content, W/CM, temperature range, age of concrete, and relative humidity. Of these factors, the type of aggregate has the greatest influence. The presence of SCMs has little or no impact on the coefficient of thermal expansion.

Chapter 9

Durability of concrete

Concrete is generally a durable material and can be expected to perform adequately for decades or even centuries in most environments, provided the following conditions are met:

- The constituent materials are suitable for the intended use.
- The concrete is proportioned to meet the exposure environment.
- The concrete is properly mixed, transported, placed, consolidated, finished, and cured.
- The quality of the concrete is appropriately controlled.
- Adequate protection is provided where required (e.g., low pH environments).
- Proper maintenance is implemented where required.

There are a number of processes that can lead to the premature deterioration of concrete, and the most common processes are listed in Table 9.1. The table categorizes the processes as being either chemical or physical, but many processes have both a chemical and a physical component. For example, physical salt attack occurs as a result of crystallization pressure or the cyclic hydration-dehydration of certain compounds, and these are really chemical processes that produce physical stresses. Many of the deterioration processes that are listed in Table 9.1 require water, and some involve the movement of species within water-filled capillaries. Thus, the rate of deterioration is controlled to some extent by the ease with which water and deleterious compounds can penetrate into and move within the pore structure of the concrete. As a consequence, the permeability of concrete is often seen as a rudimentary measure of its durability. Concrete with low permeability is generally more durable than concrete with high permeability. As a consequence, this chapter starts with a discussion on permeability.

Table 9.1 Causes of concrete deterioration

Chemical	Physical
Acid attack	Freezing and thawing
Sulfate attack	De-icer salt scaling
Delayed ettringite formation	Abrasion and erosion
Attack by other chemicals	Physical salt attack
Corrosion of embedded metals	Fire
Alkali-aggregate reactions	

Source: Based on Mindess, S., et al., *Concrete*, 2nd ed., Prentice Hall, Englewood Cliffs, NJ, 2003.

9.1 PERMEABILITY

Concrete is a porous material and fluids (liquids or gases) may flow through concrete under certain circumstances. The permeability of concrete to fluids is generally measured by applying a pressure gradient across a concrete sample and measuring the rate of fluid flow. The permeability of concrete to water, often termed the hydraulic conductivity, is often measured on a water-saturated cylindrical sample of concrete by sealing the curved surface of the sample, applying water under pressure to one flat face, and measuring the rate of water flowing out from the other face. The coefficient of water permeability is then calculated using D'Arcy's equation as follows:

$$k = \frac{Q}{A} \cdot \frac{l}{\Delta h} \tag{9.1}$$

where

k = coefficient of water permeability or hydraulic conductivity (m/s)
Q = flow rate (m^3/s)
A = cross-sectional area of sample (m^2)
l = length of the sample in the direction of flow (m)
Δh = difference in hydraulic head across the sample (m)

One of the earliest studies of the effect of supplementary cementing materials (SCMs) on water permeability was that of Davies (1954), who examined the permeability of concrete pipe and the effects of replacing 30 or 50% of the cement with Class F fly ash from two different sources (Table 9.2). The data indicate that fly ash concrete is more permeable at 28 days, especially at higher levels of replacement. However, in more mature concrete (180 days) the trend is reversed, with fly ash concrete being significantly less permeable.

Table 9.2 Effect of fly ash on the permeability of concrete

Fly ash			Relative permeability	
Source	% by mass	W/CM	28 days	6 months
—	0	0.75	100	26
Chicago	30	0.70	220	5
	60	0.65	1410	2
Cleveland	30	0.70	320	5
	60	0.65	1880	7

Source: Data from Davies, R.E., Pozzolanic Materials—With Special Reference to Their Use in Concrete Pipe, Technical Memo, American Concrete Pipe Association, 1954.

Table 9.3 Effect of silica fume on water permeability

	Coefficient of permeability ($\times 10^{-13}$ m/s)			
	7 days	28 days	91 days	182 days
100% Type V	6.3	3.8	1.3	0.3
10% silica fume	10.0	0.9	0.6	0.4
20% silica fume	6.3	<0.1	0.4	<0.1

Source: Data from Hooton, R.D., ACI Materials Journal, 90(2), 143–151, 1993.

Hooton (1993) conducted tests on paste samples with silica fume, and the data are shown in Table 9.3. Although the pastes with silica fume were more permeable at 7 days, there was a clear trend in reduced permeability, with increasing silica fume content at 28 days.

It is very difficult to accurately measure the permeability of low-permeability concrete (concrete with low W/CM and SCMs), and much of the data in the literature is limited to measurements made on paste specimens. Because of these difficulties, it is now more common to use rapid indirect measures of concrete permeability, such as ASTM C 1202, *Standard Test Method for Electrical Indication of Concrete's Ability to Resist Chloride Ion Penetration*, often referred to as the rapid chloride permeability test (RCPT). The effect of SCMs on the results of this test will be discussed in Section 9.2.3.

9.2 CORROSION OF STEEL REINFORCEMENT, CHLORIDE INGRESS, AND CARBONATION

9.2.1 Corrosion of steel reinforcement

The corrosion of embedded steel reinforcement in concrete is the predominant cause of premature failure of reinforced concrete structures

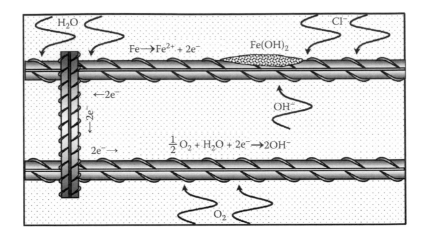

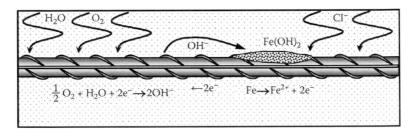

Figure 9.1 Macrocell and microcell corrosion of steel in concrete.

in North America and many other countries worldwide. The high pH of the pore solution of concrete (pH > 13.0) is actually beneficial to embedded steel, because under these conditions a dense layer of iron oxide forms on the surface of the steel and protects the steel from corrosion. This protective layer is known as the passive layer, and when it is present, the steel is said to be passivated. Unfortunately, the passive layer becomes unstable in the presence of chloride ions or when the pH of the concrete is reduced by the process known as carbonation. The ability of concrete to resist chloride ion penetration and carbonation, and the effect of SCMs on the resistance, will be discussed in Sections 9.2.2 to 9.2.4.

Once the passive layer becomes unstable and the steel is depassivated, corrosion of the steel may occur if there is sufficient oxygen and moisture in the concrete to sustain the corrosion process. Figure 9.1 shows schematics illustrating the corrosion process in concrete. Corrosion is an electrochemical process involving two half-cell reactions, one at the anode and one at the cathode. Macrocell corrosion represents the case where the anode and cathode are located on two distinct but electrically connected areas of steel that may be in different environments (e.g., top and bottom mats in a bridge deck slab). Microcell corrosion represents the case where the anodes and cathodes are intimately mixed on the same piece of steel. The anode is the location where the metallic iron oxidizes going into solution as ferrous ions; the anodic half-cell reaction is represented by the equation

$$Fe \rightarrow Fe^{2+} + 2e^- \qquad (9.2)$$

Water and oxygen are reduced at the cathode to form hydroxide ions, and the cathodic half-cell reaction may be represented by the equation

$$\tfrac{1}{2}O_2 + H_2O + 2e^- \rightarrow 2OH^- \qquad (9.3)$$

Both of these (half-cell) reactions occur simultaneously and at the same rate, with the electrons (e^-) produced by the anodic reaction being consumed by the cathodic reaction. The overall reaction can be obtained by adding Equations 9.2 and 9.3 together with the following result:

$$\tfrac{1}{2}O_2 + H_2O + Fe \rightarrow Fe^{2+} + 2OH^- \qquad (9.4)$$

The ferrous hydroxide, $Fe(OH)_2$, produced is the first corrosion product and will oxidize and hydrate further to produce a variety of iron oxides and hydroxides. These corrosion products occupy many times the volume originally occupied by the metallic iron (or steel). The overlying concrete cannot accommodate the volume increase, and eventually steel corrosion will result in cracking, spalling, and delamination of the concrete cover (see Figure 9.2). Damage to the concrete usually occurs after very little loss of the steel cross section has occurred, and in most cases, damage due to corrosion of steel is manifested long before the reinforced concrete loses load-carrying capacity. There are a number of strategies for repairing concrete damaged by corrosion of steel reinforcement. A more detailed discussion of steel corrosion in concrete and methods of prevention and repair can be found in the texts by Broomfield (1997) and Bentur et al. (1997).

A two-stage model, such as that shown in Figure 9.3 (adapted from Tuutti, 1982), is often used to represent the process of steel corrosion. If some index of damage (e.g., cracking) is plotted on the vertical axis and time on the horizontal axis, the model has two regions: (1) the initiation period, which represents the time it takes for sufficient chlorides or the carbonation front to penetrate the concrete cover and reach the steel, thereby initiating corrosion; and (2) the propagation period, which represents the time it takes for corrosion to propagate sufficiently to produce an unacceptable level of

Figure 9.2 **(See color insert.)** Cracking, spalling, and delamination of concrete cover due to corrosion of embedded reinforcement.

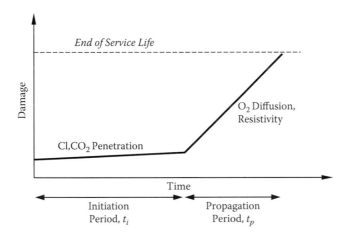

Figure 9.3 Two-stage service life model. (Proposed by Tuutti, K., *Corrosion of Steel in Concrete*, Report 4-82, Swedish Cement and Concrete Research Institute, 1982.)

damage and for the structure to reach the end of its service life. Note that the end of service life does not mean that the structure has to be replaced, but it does mean that some level of repair is necessary.

SCMs can influence the service life by changing either the initiation period or the propagation period, or both. The rate of corrosion depends predominantly on the electrical resistivity of the concrete between the anode and cathode, and the availability of oxygen and water at the cathode. Because SCMs generally increase the electrical resistivity (see later) and reduce the rate of oxygen and water transport in concrete, they may be expected to reduce the corrosion rate, and thus extend the propagation period. Testing that is designed to replicate macrocell conditions will indicate that the effect of SCMs on the corrosion rate is significant due to the distance between the anode and cathode and the high resistivity of the concrete that separates them. However, in good quality concrete microcell corrosion is most likely to be the dominant form of corrosion, and SCMs have much less influence on the microcell corrosion rate. The influence of SCMs on the initiation period is of more significance, as they can have a significant impact on the rate of chloride ingress and carbonation. This is discussed in the remainder of this section.

9.2.2 Chloride ingress

Chloride transport in concrete can occur by a number of mechanisms, and these are summarized in Table 9.4 together with a brief description of the process and test methods for evaluating the process. All of the methods listed, or variations thereof, have been used to determine the effect of SCMs on chloride ingress, and the tests invariably show that SCMs improve the resistance of concrete to chloride ion penetration, although the extent of the improvement depends on the following factors:

- The nature and amount of the SCM (or SCMs) being used
- W/CM, and to a lesser extent, the composition of the portland cement used
- The maturity of the concrete
- The exposure conditions
- The test method

The bulk diffusion test (ASTM C 1556) involves immersing a saturated disc of concrete (typically 100 mm diameter × 50 mm thick), which is sealed on all but one face, in a solution of sodium chloride. After a certain period of time (ASTM C 1556 states a minimum of 35 days) the sample is ground to produce power samples at increasing depth increments from the exposed face. The apparent chloride diffusion coefficient can then be calculated

Table 9.4 Chloride transport mechanisms and methods of test

Mechanism	Description	Test methods
Diffusion	Chloride ions move from areas of high concentration at the surface to areas of low concentration within the concrete. Process occurs in water-saturated concrete without movement of water.	ASTM C 1556: bulk diffusion test Steady-state diffusion cell
Capillary suction	Saline water is "sucked" into unsaturated concrete as a result of a moisture gradient.	AASHTO T259: 90-day salt ponding test
Migration	Negatively charged chloride ions migrate due to an electrical gradient; the chlorides are drawn toward the anode.	AASHTO TP 64: rapid chloride migration test
Convection	Saline water is driven into concrete by a hydraulic pressure gradient.	Water permeability test with saline water at the upstream face
Mixed	Specimens placed in field exposure conditions (e.g., marine exposure) or samples taken from structures exposed to chlorides.	Establish chloride profile by suitable procedure (e.g., method in ASTM C 1556)

using a numerical solution to Fick's law for non-steady-state diffusion in a semi-infinite medium, which is

$$C_x = C_s \left(1 - erf \frac{x}{2\sqrt{D_a t}} \right) \tag{9.5}$$

where

C_x = the chloride content at depth x
C_s = the chloride content at the surface ($x = 0$)
x = depth (m or in.)
t = time (s or y)
D_a = diffusion coefficient (m²/s or in.²/year)

Note that the chloride contents C_x and C_s can be expressed in any units (e.g., percent by mass, ppm, or mass per unit volume), provided they are measured in the same units. The values of C_s and D_a are found by fitting Equation 9.5 to the experimental profile (C_x-x plot) measured after the selected period, t, of immersion.

Figure 9.4 shows chloride profiles from tests conducted on cores from a pavement that was constructed with various concrete mixes having W/CM

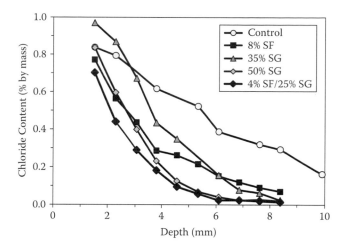

Figure 9.4 Chloride profiles for concrete (W/CM = 0.42) with varying levels of silica fume and slag. (Data from Bleszynski, R., The Performance and Durability of Concrete with Ternary Blends of Silica Fume and Blastfurnace Slag, PhD thesis, University of Toronto, 2002.)

= 0.42 and various combinations of silica fume and slag (Bleszynski et al., 2002). The cores were cut when the pavement was 2 years old, and slices from these cores were immersed in chloride solution and tested in accordance with ASTM C 1556 using a soaking period of 120 days. Table 9.5 presents the diffusion coefficients calculated from the five profiles shown in Figure 9.4, together with data from two other mixes used in the study. The benefit of using SCMs is apparent, with significantly lower chloride levels being detected at depth (Figure 9.4) and diffusion coefficients being three to four times lower than the control for the mixes with 8% silica fume or 35%

Table 9.5 Calculated diffusion coefficients ($\times 10^{-12}$ m²/s) for concrete (W/CM = 0.42) with varying levels of silica fume and slag

Silica fume content (%)	Slag content (%)			
	0	*25*	*35*	*50*
0	4.8		1.3	0.61
4		0.63		
5			0.75	
6		0.70		
8	1.4			

Source: Data from Bleszynski, R., et al., *ACI Materials Journal*, 99(5), 499–508, 2002.

slag, and six to eight times lower for the mixes with 50% slag or ternary blends of silica fume plus slag.

As with other properties, the resistance of concrete to chloride penetration increases as the concrete matures, and this is especially the case of concrete containing SCMs such as fly ash and slag. Figure 9.5 shows chloride profiles measured in concrete blocks exposed to seawater spray on the southeastern coast of the United Kingdom. (Thomas and Bamforth, 1999). The blocks were constructed with concrete mixes containing 100% portland cement, 30% fly ash, or 70% slag; the mixes were designed to achieve equivalent strength at 28 days and had W/CM = 0.66, 0.54, and 0.48, respectively. At early ages (6 months to 1 year), the extent of chloride penetration was similar for all three mixes, but after longer periods of exposure (2 years and beyond) the increased resistance to chloride penetration of the concrete containing fly ash and slag became apparent, with little continued penetration seeming to occur beyond 2 or 3 years in these concretes.

The reduction in the diffusion coefficient with time can be represented by a power-law relationship such as that shown in Equation 9.6:

$$D_t = D_{ref} \cdot \left(\frac{t_{ref}}{t} \right)^m \tag{9.6}$$

where

D_t = diffusion coefficient at time t
D_{ref} = diffusion coefficient at some reference time t_{ref}
m = decay coefficient

The value m represents the rate of decay of the diffusion coefficient increases with increasing fly ash and slag content. Thomas and Bamforth (1999) used t_{ref} = 28 days and found that the values of m and D_{28} (diffusion coefficient at the reference time of 28 days) that best fit the profiles at all six ages were as follows:

	Portland cement	Fly ash	Slag
D_{28} (m²/s)	8×10^{-12}	6×10^{-12}	25×10^{-12}
m	0.1	0.7	1.2

Figure 9.6 shows the time-dependent nature of the diffusion coefficient for the three concrete mixtures.

Other pozzolans can also be expected to reduce the diffusion coefficient of concrete, and the degree of reduction will vary, depending mainly on the nature of the pozzolan and the level of replacement. Highly reactive

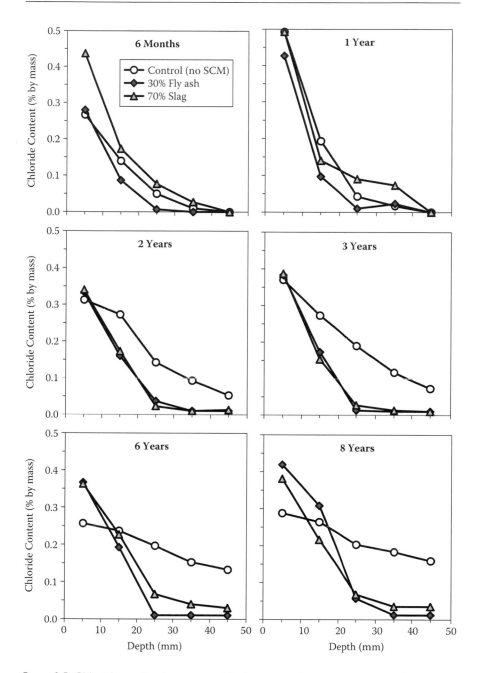

Figure 9.5 Chloride profiles in concrete blocks exposed to seawater spray. (Data from Thomas, M.D.A., and Bamforth, P.B., *Cement and Concrete Research*, 29, 487–495, 1999.)

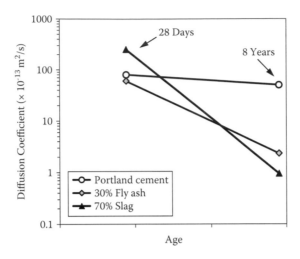

Figure 9.6 Change in diffusion coefficient with time. (Data from Thomas, M.D.A., and Bamforth, P.B., *Cement and Concrete Research*, 29, 487–495, 1999.)

pozzolans, such as metakaolin, can be expected to behave in a manner similar to that of silica fume by significantly reducing the diffusion coefficient at early ages compared to more slowly reacting pozzolans (e.g., fly ash) and slag. Figure 9.7 shows data for concrete containing high-reactivity metakaolin (Gruber et al., 2001).

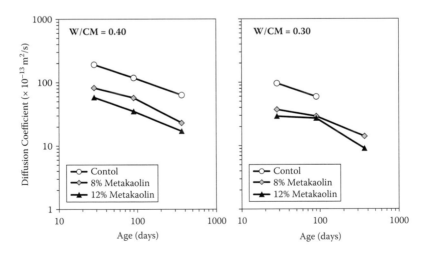

Figure 9.7 Effect of metakaolin on diffusion coefficient. (Data from Gruber, K.A., et al., *Cement and Concrete Composites*, 23(6), 479–484, 2001.)

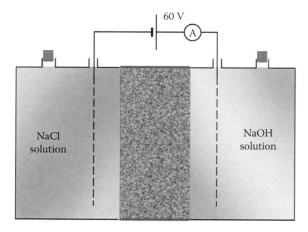

60 V

A

NaCl
solution

NaOH
solution

Figure 9.8 Schematic of RCPT (ASTM C 1202 and AASHTO T 277).

9.2.3 Rapid chloride permeability test (RCPT)

The rapid chloride permeability test (see Figure 9.8) is the commonly used name for the *Standard Test Method for Electrical Indication of Concrete's Ability to Resist Chloride Ion Penetration* (ASTM C 1202 and AASHTO T 277). In this test a saturated concrete disc, approximately 100 mm diameter × 50 mm thick (4 × 2 in.), is sandwiched between two cells, each of which contains an electrode and an electrolyte (3% NaCl as the catholyte and 0.3 M NaOH as the anolyte). A potential difference of 60 V is applied across the cell, and the resulting current is measured over a 6-hour period. The total electrical charge that passes over the 6-hour period is calculated (from the time-current data) and reported in units of Coulombs. ASTM C 1202 provides the following guidelines to interpret the results (adapted from Whiting, 1981):

Charge passed (Coulombs)	Chloride ion penetrability
>4,000	High
2000–4000	Moderate
1000–2000	Low
100–1000	Very low
<100	Negligible

It is now generally understood that this test measures neither permeability nor chloride ion penetrability, but provides a measure of the electrical conductivity of the concrete. However, despite a number of limitations,

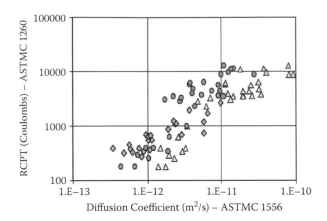

Figure 9.9 Correlation between results of rapid chloride permeability test and the bulk diffusion test determined on concrete mixtures tested at the University of Toronto and the University of New Brunswick between 1995 and 2005.

the test does correlate reasonably well with other mass transport measurements (Whiting, 1981; Stanish et al., 2001) and has the advantage of being relatively simple to conduct. Figure 9.9 shows the relationship between the results of the RCPT and a bulk diffusion test (ASTM C 1556) for a wide range of concretes tested at the University of Toronto and the University of New Brunswick between 1995 and 2005; there is a broad correlation between the results of these two tests. Since the RCPT is merely an electrical conductivity test, running the test for 6 hours is really not necessary, as the conductivity, or its reciprocal, resistivity, can be measured in a matter of minutes. Future versions of the standard test method ASTM C 1202 will include a procedure for rapidly determining the electrical resistivity of the concrete as an alternative to RCPT.

The RCPT is specified by a number of state highway departments in the United States and provincial agencies in Canada for reinforced concrete exposed to chloride ions (e.g., bridges exposed to de-icing salts or seawater). The maximum allowable RCP values vary with different agencies, but are typically in the range of 1000 to 2000 Coulombs, with the age of test varying from 28 to 56 days. As with chloride diffusion, the RCP value is significantly reduced at early ages (e.g., 28 days) by highly reactive pozzolans such as silica fume and metakaolin, but longer periods of time are required before the full benefits of fly ash and slag are realized, especially in concrete with higher levels of replacement of these materials. A study by the Virginia Transportation Research Council (Lane and Ozyildirim, 1999) demonstrated that the long-term benefits of fly ash and slag can be evaluated at 28 days using an accelerated curing regime consisting of 7 days at 23°C (73°F) and 21 days at 38°C (100°F). Figure 9.10 shows the results of

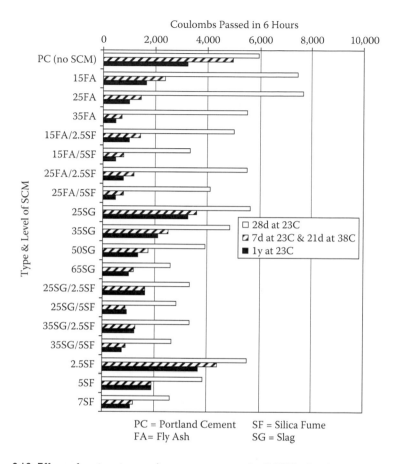

Figure 9.10 Effect of curing time and temperature on the RCPT value for concrete with varying levels of SCM. (Data from Lane, D.S., and Ozyildirim, C., *Combinations of Pozzolans and Ground, Granulated, Blast-Furnace Slag for Durable Hydraulic Cement Concrete*, VTRC 00-R1, Virginia Transportation Research Council, Charlottesville, VA, 1999.)

their study, which indicate that the 28-day accelerated curing regime gives a reasonable prediction of the RCP values obtained after curing for 1 year at 23°C (73°F).

At typical levels of replacement and at ages beyond a year or so, concrete containing fly ash or slag can be expected to be at least as resistant to chlorides as concrete containing silica fume. At even later ages (e.g., 10 years or more) fly ash and slag concrete may provide improved resistance to chlorides compared with silica fume concrete. Ternary blends of silica fume plus fly ash or silica fume plus slag have a high resistance at both early and later ages. This concept is demonstrated in Figure 9.11. Portland cement

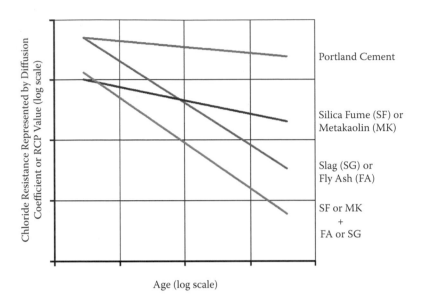

Figure 9.11 Reduction in diffusion coefficient in concrete containing one or two SCMs.

concrete has a relatively low resistance to chloride ion penetration at early age, and its resistance increases slowly with time. Concrete containing fly ash or slag may not show much improvement in terms of chloride resistance at early age, but the resistance increases rapidly with age, and at later ages both fly ash and slag can be expected to have a much improved chloride ion resistance. Highly reactive pozzolans such as silica fume and metakaolin increase the chloride resistance at early ages, and when combined with fly ash or slag in ternary cement blends, the resistance increases with age. Ternary cement blends are now favored in many cases where a high level of chloride resistance is required.

9.2.4 Carbonation

When CO_2 gas from the atmosphere penetrates into concrete it will react with the main cement hydration products. The reaction with calcium hydroxide can be written as follows:

$$CO_2 + H_2O \rightarrow CO_3^{2-} + 2H^+ \tag{9.7}$$

$$CO_3^{2-} + 2H^+ + Ca(OH)_2 \rightarrow CaCO_3 + 2H_2O \tag{9.8}$$

The overall reaction is often written as follows:

$$CO_2 + Ca(OH)_2 \rightarrow CaCO_3 + H_2O \tag{9.9}$$

However, it is important to understand that the CO_2 first has to dissolve in water to form carbonate ions, and thus that water is required; carbonation will not occur in dry concrete. Carbonate ions, CO_3^{2-}, present in some groundwaters can react directly according to Equation 9.8. Although Equation 9.9 is often the only equation that is presented to represent carbonation, the other hydrates, such as C-S-H and various aluminoferrite phases, will also react, and the cement matrix in carbonated concrete consists of calcium carbonate, silica gel, hydrated iron and aluminum oxides, and gypsum. In addition, the alkali hydroxides in the pore solution will carbonate. A small amount of shrinkage occurs as a result of carbonation, but normally the mechanical properties are not significantly affected by the process, as the resulting calcium carbonate serves as a binder. The main consequence of carbonation is that the pH of the concrete drops from above pH 13 to somewhere in the region of pH 8. If the carbonation front reaches the embedded steel reinforcement, the loss of alkalinity will result in the loss of the protective passive layer and corrosion of the steel will be initiated. If there is sufficient oxygen and water available in the concrete the steel will corrode, eventually leading to damage to the concrete.

The depth of carbonation in concrete can be readily detected by spraying a freshly fractured surface with phenolphthalein solution, which is a pH indicator that turns purple when pH > 9.2 and remains colorless when pH < 9.2 (see Figure 9.12).

The rate of carbonation of properly proportioned and well-cured concrete is slow, and provided adequate cover is provided to protect the embedded steel, carbonation-induced corrosion is unlikely to occur during the typical service life of a reinforced concrete structure. When corrosion damage due to carbonation does occur in concrete structures, this is usually a result of a combination of poor quality concrete, inadequate curing, and insufficient concrete cover.

The carbonation rate is often represented by a power-law relationship as follows:

$$d_c = k \cdot t^m \tag{9.10}$$

where
    d_c = depth of carbonation (mm or in.)
    k = carbonation coefficient (mm/year$^{\frac{1}{2}}$ or in./year$^{\frac{1}{2}}$)when $m = 0.5$
    t = time (year)
    m = constant (typically a value of $m = 0.5$ is used)

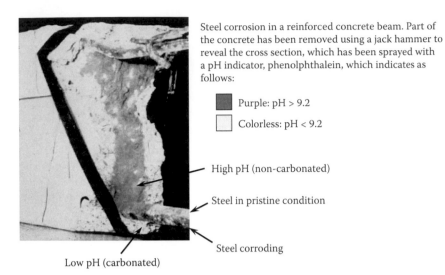

Steel corrosion in a reinforced concrete beam. Part of the concrete has been removed using a jack hammer to reveal the cross section, which has been sprayed with a pH indicator, phenolphthalein, which indicates as follows:

■ Purple: pH > 9.2

□ Colorless: pH < 9.2

— High pH (non-carbonated)

← Steel in pristine condition

← Steel corroding

Low pH (carbonated)

Figure 9.12 **(See color insert.)** Carbonation-induced steel corrosion in a reinforced concrete beam.

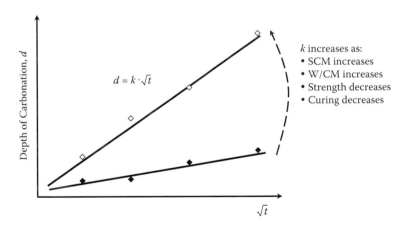

Depth of Carbonation, d

$d = k \cdot \sqrt{t}$

k increases as:
• SCM increases
• W/CM increases
• Strength decreases
• Curing decreases

$\sqrt{t}$

Figure 9.13 Rate of carbonation of concrete.

The value of the carbonation coefficient, k, depends on the following (Figure 9.13):

- Strength and W/CM: k increases as strength decreases or W/CM increases.
- Curing: k increases as the duration of curing decreases.
- SCM content: k increases as the level of SCM increases.

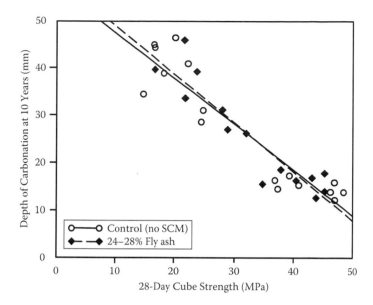

Figure 9.14 Relationship between depth of carbonation and strength for concrete with and without fly ash. (Data from Matthews, J.D., in *Ashtech '84*, London, 1984, p. 398A.)

- Exposure: Carbonation occurs more rapidly when the relative humidity (RH) is in the range of 45 to 65%. At higher RH, the penetration of CO_2 is hindered by water partially filling the pores in the concrete. At lower RH, Equation 9.7 is limited by the availability of moisture. Note that although the maximum rate of carbonation occurs when 45% < RH < 65%, there is insufficient moisture available for corrosion unless RH > 80% in the vicinity of the steel.

An extensive review of carbonation studies by Parrott (1987) concluded that for a given water-to-binder ratio (W/CM) the rate of carbonation increases as the amount of SCM increases. This presumably is due to the reduced amount of $Ca(OH)_2$ in concrete containing pozzolans or slag. Much of the data available indicates that the rates of carbonation are similar for concretes, with or without SCM, when proportioned to achieve the same 28-day compressive strength.

Figure 9.14 shows the relationship between the 28-day compressive strength and the depth of carbonation for 10-year-old concrete with and without fly ash (Matthews, 1984), and a similar relationship is shown in Figure 9.15 for 9-year-old concrete with slag (Osborne, 1986). For a given compressive strength the rate of carbonation appears to be similar for concretes regardless of the level of fly ash or slag. However, in these studies concretes of the same strength had decreasing W/CM with increasing fly

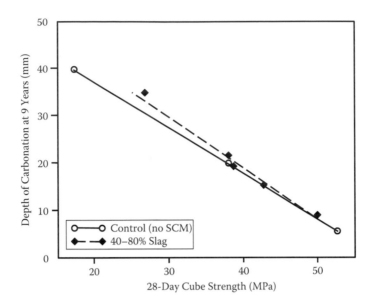

Figure 9.15 Effect of slag on the relationship between 28-day compressive strength and depth of carbonation at 9 years. (Data from Lewis, R., et al., in *Advanced Concrete Technology: Constituent Materials*, ed. J. Newman and B.S. Choo, Elsevier, Oxford, 2003, pp. 3/1–3/66; Osborne, G.J., *Durability of Building Materials*, 4, 81–96, 1986.)

ash and slag contents, which means that if the concretes were compared on the basis of equal W/CM, there would be increasing carbonation with increasing levels of fly ash and slag.

There are few studies that suggest that somewhat higher carbonation rates may occur in concrete containing fly ash and slag compared with similar strength portland cement concrete, especially in low-strength, poorly cured concrete with relatively high levels of replacement. Figure 9.16 shows the depth of carbonation for concretes stored outdoors but sheltered from rain for 10 years as a function of strength, fly ash content, and the duration of moist curing. Table 9.6 shows carbonation rate coefficients calculated from these measured depths. As expected, there is an increase in carbonation with decreasing strength, reducing moist curing and increasing fly ash content. The differences between concretes with 0 to 30% fly ash are generally small, but the poorly cured lower-strength concrete with 50% fly ash carbonated to a significantly greater extent. A study on the effects of slag on the carbonation of field structures (Osborne, 1989) showed that the depth of carbonation depended mainly on the slag content and the microclimate. In dry, sheltered environments the depth of carbonation increased significantly with slag content, especially at 70% slag, and Osborne (1989) recommended that the slag replacement level be restricted to 50% for slender reinforced concrete

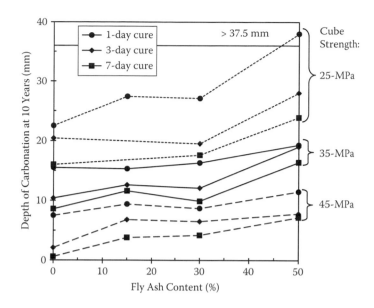

Figure 9.16 Effect of strength, curing, and fly ash content on carbonation of concrete stored in an outdoor, sheltered environment of 10 years. (Data from Thomas, M.D.A., et al., in *Proceedings of the 4th ACI/CANMET International Conference on the Durability of Concrete*, ed. V.M. Malhotra, ACI SP-192, Vol. 1, 2000, pp. 539–556.)

Table 9.6 Effect of strength, curing, and fly ash on carbonation rate coefficient, k (mm/year$^{1/2}$)

Strength grade[a] MPa (psi)	Moist curing (days)	Fly ash content (%)			
		0	15	30	50
18 (2610)	1	6.8	8.3	8.3	10.2
	3	5.8	6.3	6.2	8.4
	7	4.9	4.6	5.3	7.1
25 (3625)	1	4.9	4.9	5.2	6.4
	3	3.1	3.7	3.9	5.9
	7	2.4	3.3	3.2	4.8
32 (4640)	1	2.4	2.8	3.0	3.9
	3	0.7	2.1	2.0	2.6
	7	0.3	1.3	1.2	1.9

Source: Data from Thomas, M.D.A., et al., in *Proceedings of the 4th ACI/CANMET International Conference on the Durability of Concrete*, ed. V.M. Malhotra, ACI SP-192, 2000, Vol. 1, pp. 539–556.

[a] Strength grade based on characteristic 28-day cylinder strengths (converted from original cube strengths).

elements exposed to such conditions. However, higher levels of slag can be used when the concrete is exposed to moisture during service and where chemical resistance to chlorides, sulfates, and seawater is required.

A recent study reported carbonation data for high-volume fly ash (HVFA) concrete (Bouzoubaa et al., 2006). In this study the maximum carbonation coefficient for concrete with 56% fly ash and W/CM = 0.32 when moist cured for 7 days was 5.04 mm/year$^{1/2}$ for indoor exposure and 2.51 mm/year$^{1/2}$ for unprotected outdoor exposure. This compares with 1.14 and 0 mm/year$^{1/2}$ (for instance, no measurable carbonation at 7 years) for portland cement concrete with the same W/CM. This indicates that the use of high levels of fly ash resulted in much increased carbonation rates. However, it was concluded by this study that carbonation is not an issue for well-cured HVFA concrete based on the calculated time to corrosion (>200 years) for reinforcing steel with a depth of cover of 40 mm (1.5 in.) in concrete exposed outdoors.

The conclusion of Bouzoubaa and coworkers (2006) is only valid if the following conditions are met: (1) high-volume fly ash concrete is proportioned with a very low W/CM (≤0.32), (2) concrete is moist cured for at least 7 days, (3) concrete is directly exposed to moisture during service (for instance, not protected from precipitation), and (4) the specified minimum cover requirements (for example, 40 mm [1.5 in.]) are met. If there are changes in the mixture proportions, if the concrete is not directly exposed to moisture (for example, it is protected from rainfall), and if 7 days' curing and 40 mm cover are specified, but not achieved in practice, the time to corrosion will be reduced substantially. For example, the data in Table 9.6 show that 25 MPa (3625 psi) concrete with 50% fly ash (W/CM = 0.41 for this mix) has a carbonation coefficient of k = 5.85 mm/year$^{1/2}$ if it is only moist cured for 3 days. If the cover actually achieved in practice is only 30 mm (1.2 in.), corrosion of the steel will be initiated in just 26 years.

Figure 9.17 shows the depth of carbonation after 1 day of curing and 6 years of storage at 20°C and 50% RH of concretes with and without silica fume plotted against the 28-day compressive strength (Sellevold and Nilsen, 1987). For a given strength, there is increased carbonation in the concretes with 20% silica fume, and this increase is enhanced at lower strengths. However, the differences between concretes with 0 to 10% silica fume are insignificant.

Some of the conflicting data in the literature are due to differences in the curing and exposure conditions used in carbonation studies. Drawing conclusions from short-term tests (<5 years or so) may result in an overestimate of the negative impact of fly ash or slag on the rate of carbonation as shown in Figure 9.18 (Hobbs, 1986). Studies comparing fly ash concretes at ages of 1 to 3 years indicate that the depth of carbonation increases with fly ash content for concretes of the same compressive strength. However, comparisons made at later ages (10 to 15 years) show that concretes of the same 28-day strength show similar performance regardless of fly ash

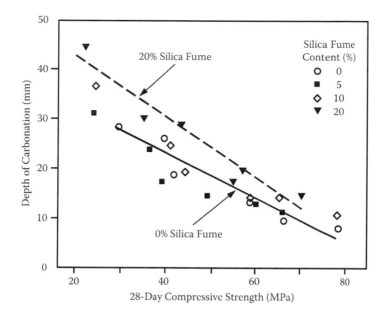

Figure 9.17 Effect of silica fume content on the relationship between strength and carbonation of concrete at 6 years. (Data from Sellevold, E.J., and Nilsen, T., in *Supplementary Cementing Materials for Concrete*, ed. V.M. Malhotra, CANMET, Ottawa, 1987, pp. 165–243.)

content. Studies that employ enriched CO_2 environments to accelerate the rate of carbonation will also tend to overestimate the extent to which SCMs increase the rate of carbonation (Thomas et al., 2000).

Based on the data available at this time, the following recommendations can be made regarding SCMs and carbonation:

- Carbonation-induced corrosion is unlikely to be a problem in properly proportioned and cured concrete that is exposed to moisture during service (providing there is adequate cover to the steel).
- Carbonation-induced corrosion is unlikely to be a problem in concrete that is in a dry condition (e.g., internal elements) where the relative humidity does not exceed 80%.
- In exposure conditions that are likely to lead to carbonation-induced corrosion, such as prolonged dry periods with occasional exposure to moisture, care should be taken to ensure that adequate cover is provided and the concrete is of adequate strength and is properly cured. If relatively high replacement levels of fly ash (e.g., 50%) or slag (e.g., 70%) are to be used, it may be prudent to provide additional cover, extend the period of initial moist curing, increase the concrete strength, or employ some combination of these strategies.

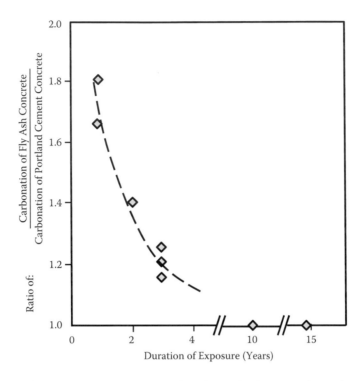

Figure 9.18 Effect of exposure period on the relationship between the carbonation of concrete with and without fly ash. (Data from Hobbs, D.W., *Advances in Cement Research*, 2, 9–13, 15–20, 1986.)

9.3 RESISTANCE TO FREEZING AND THAWING, AND DE-ICER SALT SCALING

Concrete in a critically saturated condition (>91.7% of the porosity filled with water) can be damaged if it is exposed to multiple cycles of freezing and thawing. The damage is generally attributed[1] to the hydraulic pressures that are formed as the water in the larger capillaries expands (by up to 9%) on freezing and forces unfrozen water into the smaller pores. If these hydraulic pressures exceed the tensile strength of the concrete, the concrete cracks. It has long been known that the hydraulic pressures can be relieved by entraining air bubbles into the concrete, and it is now generally established that concrete will be resistant to cyclic freezing and thawing provided the following conditions are met:

- The concrete is adequately air entrained.

---

[1] The hydraulic pressure theory was proposed by Powers (1945). There have since been more complex theories and the reader is referred to the text by Pigeon and Pleau (1995) for a more detailed treatment of freeze-thaw mechanisms.

- The aggregate is frost resistant.
- Adequate strength is achieved prior to the first freeze (>3.5 MPa or 500 psi).
- The concrete is of adequate strength to resist cyclic freezing and thawing (>28 MPa or 4000 psi).

These four conditions are generally applicable regardless of whether the concrete contains supplementary cementing materials. There have been studies that indicate that the freeze-thaw durability of low-strength or "lean" concretes is impaired by the use of SCMs such as fly ash (e.g., Whiting, 1989), but providing the concrete is properly proportioned (e.g., W/CM ≤ 0.50), concrete containing SCMs can be expected to be durable when exposed to cycles of freeze-thaw in a saturated condition.[1]

There are conflicting data concerning the freeze-thaw resistance of non-air-entrained concrete containing silica fume. A number of studies have shown that such concrete can provide adequate freeze-thaw resistance when the W/CM is below 0.35, and this is possibly due to the internal self-desiccation of the concrete, resulting in it being less than critically saturated. However, there are studies that indicate that even at extremely low W/CM, non-air-entrained concrete with silica fume does not have adequate freeze-thaw resistance (Malhotra et al., 1987). It is recommended that all concrete be air entrained if it is to be exposed to freezing and thawing while in a critically saturated condition.

The influence of SCMs on the resistance of concrete to freezing and thawing in the presence of de-icing chemicals, such as rock salt (NaCl), is less clear. Concrete, even when air entrained, exposed to these conditions can, under certain circumstances, suffer deterioration in the form of surface scaling, which describes the progressive raveling of material from the surface; examples of this form of deterioration are shown in Figure 9.19. The process is often referred to as de-icer salt scaling, salt scaling, or just scaling. This form of concrete deterioration is largely confined to hand-finished flatwork such as concrete driveways and sidewalks, and is rarely seen on slipformed (e.g., pavements or bridge decks) or formed surfaces.

There are a number of reasons why the application of de-icing salts increases the severity of freeze-thaw attack at the exposed surface of concrete:

1. Increased number of freeze-thaw cycles.
2. Salts increase the degree of saturation.
   - NaCl is hygroscopic.
   - The degree of saturation at a given temperature and relative humidity is higher in salt-contaminated pore water.

---

[1] This statement refers to concrete exposed to freezing and thawing in the absence of de-icing or anti-icing chemicals.

Figure 9.19 **(See color insert.)** Examples of de-icer salt scaling of concrete driveway (left) and sidewalk (right).

3. Salts encourage a differential response of layers to freezing.
   • Salt concentration gradient causes differences in freezing behavior of adjacent layers. Differential stresses may be caused as a result of layer-by-layer freezing.
4. Application of salts subjects concrete to thermal shock.
   • Melting ice or snow requires energy that is drawn from the concrete and causes a rapid temperature drop of the concrete at the surface.
5. Salts exacerbate osmotic pressures.
   • As the water freezes the salt concentration in the unfrozen water will be greater in concrete exposed to de-icing salts, leading to an increased concentration gradient and resulting osmotic pressures.

Laboratory scaling tests are performed by freeze-thaw cycling concrete slabs ponded with solutions of de-icing salts (see Figure 9.20). The ASTM C 672 standard test method specifies a 4% $CaCl_2$ solution be ponded on the surface of the slab and a visual rating to determine the performance of the concrete (Figure 9.20). Variations on this test include the use of different salt solutions (e.g., 3% NaCl) and measurements to be made of the mass of material that scales from the slab during test. There are a number of publications that show that concrete containing SCM is less resistant to de-icing salt scaling when tested using accelerated laboratory tests of this nature. Figures 9.21 and 9.22 show examples of such test data for slag and fly ash, respectively, both studies indicating that the incorporation of either of these materials renders the concrete more susceptible to scaling. However, other studies have shown generally acceptable performance of concrete in accelerated tests provided the W/CM ≤ 0.45 and the level of SCM is kept reasonable (e.g., ≤25% fly ash, ≤40% slag, and ≤10% silica fume).

The ACI 318-08 Building Code limits the amount of SCM that can be used in concrete exposed to de-icing salts; these limits are shown in Table 9.7. Unfortunately, the limits apply to all concrete and make no distinction between formed and finished surfaces.

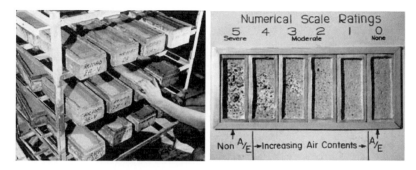

Figure 9.20 **(See color insert.)** Laboratory salt scaling tests showing concrete slabs with dyke to contain salt solution (left) and visual rating system (right). (From Caldarone, M.A., et al., *Guide Specification for High-Performance Concrete for Bridges*, EB233, 1st ed., Portland Cement Association, Skokie, IL, 2005; Kosmatka, S.H., and Wilson, M.L., *Design and Control of Concrete Mixtures*, EB001, 15th ed., Portland Cement Association, Skokie, IL, 2011. Printed with permission from the Portland Cement Association.)

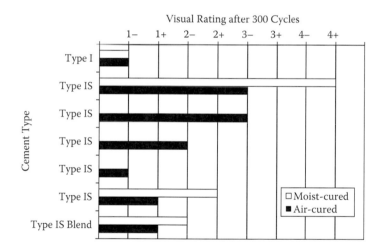

Figure 9.21 Results of de-icer salt scaling tests conducted on concrete (W/CM = 0.40) produced with Type IS portland blast furnace slag cement. (Data from Klieger, P., and Isberner, A.W., *Journal, PCA Research and Development Department Laboratories*, 9(3), 2–22, 1967.)

Despite the overall inferior performance shown by SCM concrete in accelerated laboratory de-icer salt scaling tests, concrete has shown satisfactory performance in the field even with levels of SCM in excess of the limits in ACI 318-08, and it has been suggested that the laboratory tests are overly aggressive and do not correlate well with field performance (Thomas, 1997). There have been a number of opportunities to compare laboratory test results with field performance. For example, Figure 9.23

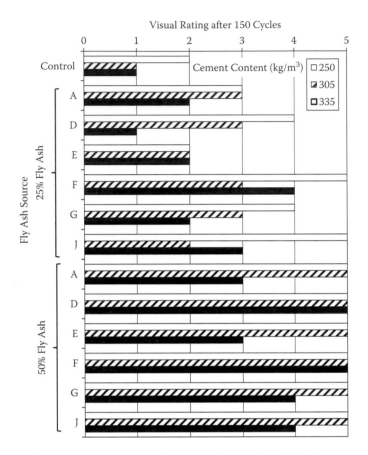

Figure 9.22 Effect of cementing materials content (strength) and fly ash content on the scaling resistance of concrete. (From Whiting, D., *Strength and Durability of Residential Concretes Containing Fly Ash*, PCA R&D Bulletin RD099, Portland Cement Association, Skokie, IL, 1989.)

shows the excellent condition of a slipformed concrete pavement with 50% Class C fly ash after salt applications over six winters, whereas concrete slabs cast from the same mix during construction exhibited a visual rating of 4 (moderate to severe scaling; see Figure 9.20) when tested in the laboratory (Naik et al., 1995). Figure 9.24 shows the construction of a number of pavement test slabs and test results for laboratory scaling tests conducted on concretes containing silica fume, slag, or combinations thereof. All of the concrete containing SCM showed inferior scaling resistance to the control mix with no SCM, but after 9 years of exposure in the field only the slab with 50% slag exhibited more scaling than the control. Details of the construction and performance of these slabs can be found elsewhere (Bleszynski et al., 2002; Hooton et al., 2008).

Table 9.7 Limits on SCM contents in concrete exposed to
de-icing salts (ACI 318-08)

Supplementary cementing material, SCM[a]	Maximum percentage of total cementing materials[b] (%)
Fly ash and natural pozzolans	25
Slag	50
Silica fume	10
Total of all SCM	50[c]
Total of fly ash, silica fume, or other pozzolans	35[c]

[a] Includes portion of SCM in blended cements.
[b] Total cementing materials include the summation of the portland cements, blended cements, fly ash, silica fume, slag, and natural pozzolans.
[c] Silica fume shall not constitute more than 10%, and fly ash or other pozzolans shall not constitute more than 25% of the total cementing materials.

Figure 9.23 **(See color insert.)** The author examines a slipformed concrete pavement with 50% Class C fly ash after exposure to de-icing salts over six winters.

Thomas (1997) conducted a survey of fly ash concrete structures in Ontario, Alberta, Michigan, Minnesota, and Wisconsin, including pavements, sidewalks, bridge decks, driveways, and barrier walls, all of which had been exposed to freezing and thawing in the presence of de-icing salts. The concretes contained up to 70% fly ash, and in most cases the concrete was determined to be in excellent condition (Figures 9.25 and 9.26), with less than excellent performance only being observed in some pavements with 50 to 60% Class F fly ash or 70% Class C fly ash (Figure 9.27). Severe scaling was observed in concrete paving with 40% Class F fly ash in a parking lot in Wisconsin (Figure 9.28). The severe scaling was observed in alternate sections of the pavement, with the sections between showing excellent performance. Inquiries made 4 years after construction revealed

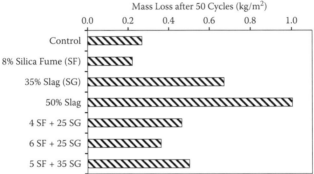

Figure 9.24 **(See color insert.)** Construction and performance of paving slabs with combinations of portland cement (PC), silica fume (SF), and slag (SG). Note excellent field performance of concrete with 5% silica fume and 35% slag after nine winters (photo top right), but inferior performance in the laboratory salt scaling test (bottom).

that the concrete mixes contained the same materials and proportions (including 40% Class F fly ash), but that the concrete sections in good condition were placed on the first day using a slipform paving machine, whereas the concrete in poor condition was poured between the previously slipformed sections approximately 7 days later and was consolidated and finished using a vibrating screed. It has been suggested that the concrete mix was designed to be placed by machine (i.e., stiff low-slump mix) and that water was added during mixing and placing to enable it to be consolidated and finished by hand on the second day. Although this is conjecture, it does serve to demonstrate that salt scaling resistance is not an inherent property of a concrete mixture, as the durability of the surface is also a function of construction practices.

The field performance of HVFA concrete (>50% fly ash) in Canada with regards to de-icer salt scaling is varied. HVFA concrete demonstration projects in Halifax included trial sections of two sidewalks, one placed in 1990 and the other in 1994 (Langley and Leaman, 1998). The first placement had

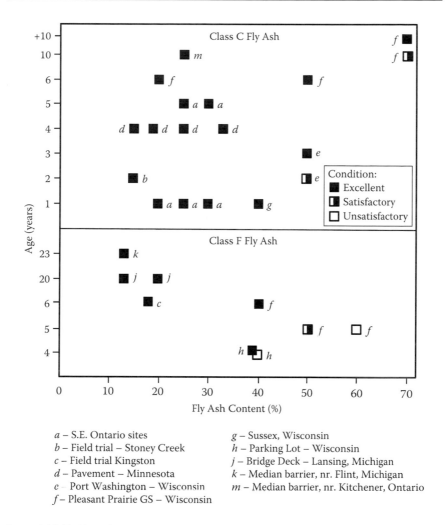

a – S.E. Ontario sites
b – Field trial – Stoney Creek
c – Field trial Kingston
d – Pavement – Minnesota
e – Port Washington – Wisconsin
f – Pleasant Prairie GS – Wisconsin

g – Sussex, Wisconsin
h – Parking Lot – Wisconsin
j – Bridge Deck – Lansing, Michigan
k – Median barrier, nr. Flint, Michigan
m – Median barrier, nr. Kitchener, Ontario

Figure 9.25 Field performance of fly ash concrete exposed to freezing and thawing in the presence of de-icing salts. (Data from Thomas, M.D.A., in *Frost Resistance of Concrete*, ed. M.J. Setzer and R. Auberg, Essen, Germany, 1997.)

a cementitious content of 390 kg/m³ (657 lb/yd³) and contained 55% of low-calcium Class F fly ash, and in 1998 it was reported (Langley and Leaman, 1998) to have shown excellent performance that was at least equivalent to the surrounding concrete. The second placement had a cement content of 400 kg/m³ (674 lb/yd³), a fly ash content of 55% (Class F), and a water content of 110 kg/m³ (185 lb/yd³), resulting in W/CM = 0.275. Figure 9.29 shows the visual appearance of the control and HVFA concrete mix in the summer of 2006, after 12 winters. The HVFA concrete has scaled heavily

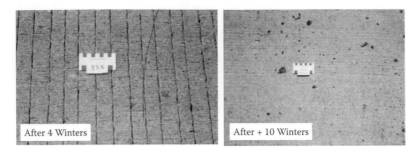

Figure 9.26 **(See color insert.)** Excellent performance of slipformed concrete pavement with 33% Class C fly ash on Highway 52 in Minnesota (left) and with 70% Class C fly ash on a Haul Road at Pleasant Prairie Generating Station in Wisconsin (right).

Figure 9.27 **(See color insert.)** Satisfactory performance (moderate scaling) in slipformed pavement with 70% Class C fly ash (left) and 50% Class F fly ash (right) at Pleasant Prairie Generating Station in Wisconsin.

Figure 9.28 **(See color insert.)** Adjacent concrete slabs on paved parking lot in Wisconsin; both slabs contain 40% Class F fly ash. Slab to left placed and finished by slipform paving machine. Slab to right placed and finished by hand.

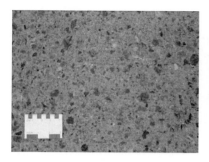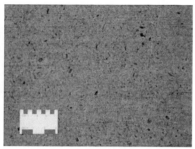

Figure 9.29 **(See color insert.)** High-volume fly ash concrete (left) and portland cement concrete (right) in Halifax after 12 years with approximately 100 freeze-thaw cycles per annum.

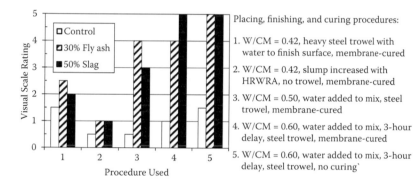

Figure 9.30 Scaling resistance of concrete with fly ash and slag—effect of poor practices.

(significantly more so than the control) but is still serviceable. It should be noted that this concrete receives frequent applications of de-icer salt and is exposed to more than 100 cycles of freeze-thaw per year (Malhotra and Mehta, 2005).

The impact of poor construction practices on the scaling resistance of concrete with and without SCM is illustrated in Figure 9.30, which shows the visual rating after 50 cycles of a salt scaling test (similar to ASTM C 672 but with 3% NaCl instead of 4% $CaCl_2$) of concrete slabs from three different concrete mixes. The mixes all started with an initial W/CM of 0.42 and a slump of approximately 25 mm (1 in.); one mix contained 30% Class F fly ash, the second 50% slag, and the third was a control mix without SCM. Ten slabs were produced from each mix using the following placing, finishing, and curing procedures:

1. Slab cast with low-slump concrete, heavy (steel) troweling, and water spray used to finish the surface; a curing compound then applied

2. Portion of the concrete transferred to a smaller mixer and a high-range water-reducing admixture added to the mix to produce a 150-mm (6-in.) slump (final W/CM = 0.42), surface struck with wood board and textured with a broom, and a curing compound applied

3. Water added to the mixer to provide a 150-mm (6-in.) slump (final W/CM = 0.50), concrete finished with steel trowel, and a curing compound applied

4. Placement delayed by 3 hours, water periodically added to the mixer to maintain 150-mm (6-in.) slump (final W/CM = 0.60), concrete finished with steel trowel, and a curing compound applied

5. As for previous (procedure 4), but no curing compound used

The results in Figure 9.30 are that both the fly ash and slag concrete showed good scaling resistance provided procedure 2 was followed. However, when improper practices, such as the addition of water, overfinishing with a steel trowel, extended delays, or inadequate curing, were used the scaling resistance was reduced significantly. The control concrete with no SCM was less sensitive to improper practices.

In summary, concrete that is resistant to de-icer salt scaling can be produced with supplementary cementing materials provided the concrete is appropriately proportioned and proper construction practices are adhered to with regards to placing, finishing, and curing the concrete. Concrete with high levels of SCM is more sensitive to improper practices, and if it is not possible to ensure that the concrete is placed, finished, and cured properly, it may be prudent to reduce the level of SCM to the maximum limits prescribed in ACI 318 for hard finished flatwork such as sidewalks and driveways.

9.4 ALKALI-SILICA REACTION (ASR)

There are two types of alkali-aggregate reaction (AAR) currently recognized in concrete: alkali-silica reaction (ASR) and alkali-carbonate reaction (ACR). Only ASR and the prevention of damaging ASR using supplementary cementing materials will be discussed in this text. ACR is not as widespread as ASR, and the preventive measures used to control ASR are generally *not* effective for ACR. Aggregates that have the potential to cause deleterious ACR should *not* be used in concrete.

Alkali-silica reaction is a reaction between the alkali hydroxides in the pore solution of concrete and certain silica minerals present in some aggregates that can, under some circumstances, produce deleterious expansion and cracking of concrete (Figure 9.31). Portland cement is the predominant source of alkalis (Na and K) in concrete, but other contributing sources may include supplementary cementing materials, admixtures, aggregates, and external sources such as de-icing salts and seawater. As discussed in

Figure 9.31 **(See color insert.)** Cracking of concrete due to ASR in (a) bridge abutment, (b) hydraulic dam, (c) retaining wall, (d) pavement, (e) bridge piers and beams, and (f) curb and gutter.

Chapter 3, the alkalis in portland cement are soluble, and beyond an age of 24 hours, the pore solution of concrete is composed almost entirely of sodium and potassium cations (Na^+ and K^+) balanced by hydroxyl anions (OH^-). Certain silica minerals present in some rock types may be attacked at high OH^- concentrations and go into solution, where they react with the alkalis to form an alkali-silica gel (which contains minor amounts of calcium). The gel is hygroscopic and will absorb water from the surrounding concrete and swell, causing the concrete to expand. When the pressure created by the swelling gel exceeds the tensile strength of the concrete, the concrete will crack. A schematic showing the process is shown in Figure 9.32. Figure 9.33 shows a thin section prepared from concrete affected by ASR. The section contains a chert particle that has reacted, resulting in expansion

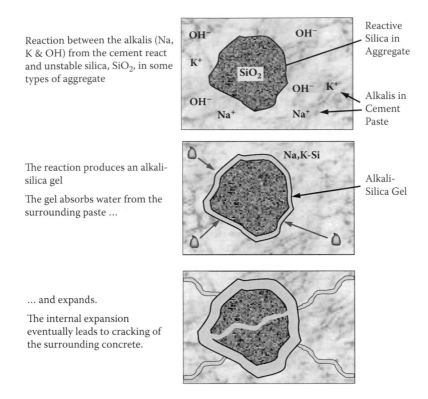

Reaction between the alkalis (Na, K & OH) from the cement react and unstable silica, SiO_2, in some types of aggregate

The reaction produces an alkali-silica gel

The gel absorbs water from the surrounding paste ...

... and expands.

The internal expansion eventually leads to cracking of the surrounding concrete.

Figure 9.32 **(See color insert.)** Mechanisms of ASR.

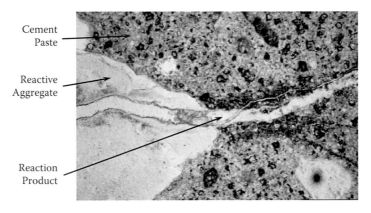

Figure 9.33 **(See color insert.)** Thin section showing site of expansive reaction in ASR-affected concrete (width of view approximately 2-mm).

as seen by the crack emanating from the chert particle into the surrounding cement paste; alkali-silica gel can be seen partially filling the crack. The chert particle is a source of expansive reaction and a reliably diagnostic symptom of deleterious ASR.

Silica, SiO_2, is a component of many, if not most, rocks. However, not all forms of silica react significantly with the pore solution of concrete. For example, quartz is a stable silica mineral owing to the fact that it has a well-ordered crystalline structure. Opal, on the other hand, has a more disordered (amorphous) structure, despite having the same chemical composition as quartz (both minerals are predominantly SiO_2). ASR will occur rapidly in concrete produced with an aggregate containing opaline silica, provided there is sufficient alkali present in the concrete. The following silica minerals are known to react deleteriously in concrete: opal, tridymite, cristobalite, volcanic glass, chert, cryptocrystalline (or microcrystalline) quartz, and strained quartz. These minerals may be found in many rock types, including shale, sandstone, silicified carbonate rocks, chert, flint, quartzite, quartz-arenite, gneiss, argillite, granite, greywacke, siltstone, arenite, arkose, and hornfels. However, this does not mean that all sources of such rocks will produce a deleterious reaction when used in concrete. For example, granitic aggregate is widely used in concrete, and only certain sources of such aggregate produce damaging ASR. The reactivity of a rock depends on the type, quantity, and habitat of the reactive minerals present. The presence of reactive minerals can usually be detected by a trained petrographer. However, appropriate performance testing of specific aggregate sources is recommended to confirm alkali-silica reactivity; such testing involves measuring the expansion of mortar bars or concrete produced with the aggregate being considered.

Methods for preventing damaging alkali-silica reaction include (1) avoiding the use of reactive aggregates, (2) minimizing the alkali content of the concrete, (3) using certain (lithium-containing) chemical admixtures, and (4) using supplementary cementing materials. Only the last of these methods, using SCM, is discussed in this text.

There are two standard test methods commonly used in North America for evaluating the effect of pozzolans or slag on expansion due to alkali-silica reaction: the concrete prism test (CPT) and the accelerated mortar bar test (AMBT). Detailed procedures for these tests can be found in ASTM C 1293 and ASTM C 1567. In the CPT, concrete prisms are stored over water in sealed containers at 38°C (100°F), and the length change is monitored over time. An aggregate is considered to be reactive if the expansion in this test (without SCM) exceeds 0.040% at 1 year. A preventive measure such as an SCM is deemed to be effective in controlling ASR with this aggregate if the expansion of the concrete incorporating SCM is less than 0.040% at 2 years. The AMBT is an accelerated test that involves measuring the length change of mortar

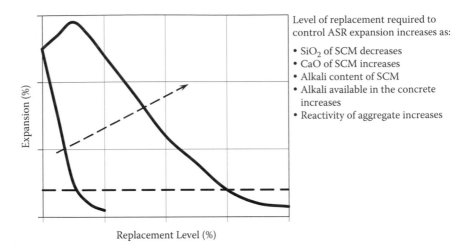

Figure 9.34 Controlling expansion with supplementary cementing materials.

bars stored in 1 M NaOH solution at 80°C (176°F). In this test, an SCM is considered to be effective in controlling deleterious expansion with a reactive aggregate if the expansion after 14 days' immersion in the solution is less than 0.10% at 14 days. Some agencies have more onerous requirements, such as an expansion limit of just 0.08% after an immersion period of 28 days.

The AMBT is a more rapid test, but the CPT is generally considered to more reliable, and the data from this test should override the data from the accelerated test when both are available. The majority of data reported in this text has been produced using the CPT.

If a series of concrete prism tests are run with a single reactive aggregate but a range of SCM replacement levels, and the 2-year expansion was plotted against the SCM level, then the data would generally resemble those shown in Figure 9.34. SCMs that are very efficient in controlling expansion would only need to be used at relatively low replacement levels to limit the expansion to below 0.040%. An example would be silica fume, which is effective when combined with high-alkali cement and most reactive aggregates, when it is used at a replacement level of 12%. SCMs that are less efficient, such as high-calcium Class C fly ashes, may be required at much higher replacement levels, such as 40 to 60%. Such materials may, in some circumstances, cause an increase in expansion compared to concrete without SCM if they are used at low levels of replacement. The efficiency of an SCM is related to its composition, particularly its CaO, SiO_2, and alkali (Na_2Oe) content; the higher the SiO_2 and the lower the CaO and Na_2Oe, the more efficient the SCM. So SCMs low in silica but high in calcium or alkali have to be used at much higher levels of replacement than SCMs

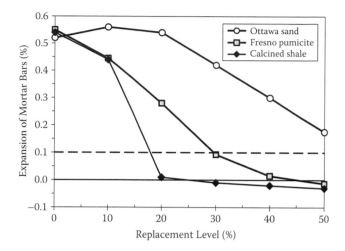

Figure 9.35 Effect of pozzolans on ASR expansion of mortars. (Data from Stanton, T.E., *Proceedings of the American Society of Civil Engineers*, 66(10), 1781–1811, 1940; Stanton, T.E., in *Pozzolanic Materials in Mortars and Concretes*, ASTM STP 99, American Society for Testing and Materials, Philadelphia, pp. 178–203, 1952.)

high in silica and low in calcium and alkali. The amount of SCM required also increases with the reactivity of the aggregate and the amount of alkali available in the concrete.

The ability of pozzolans to control ASR was recognized by Stanton (1940) in his landmark paper that first described the reaction as a cause of concrete deterioration. In a later paper (Stanton, 1952) he showed that two types of pozzolans, a pumicite and a calcined shale, could effectively eliminate damaging reaction in mortar bars provided they were used at a sufficient level of replacement (see Figure 9.35). He also conducted tests using ground Ottawa sand as a replacement to determine the effect of diluting the cement with an "inert" material. As shown in Figure 9.35, the beneficial effect of the pozzolans is greater than that expected from an inert diluent.

In the years that followed Stanton's joint discovery of ASR and the ability of pozzolans to control expansion, there was a considerable amount of work to confirm his findings and to investigate the effect of other materials, including fly ash and slag. Much of this work was conducted by the U.S. Corps of Engineers and the U.S. Bureau of Reclamation, and the findings led to natural pozzolans being used to limit the risk of ASR in concrete structures such as those of the Davis Dam, which used a calcined shale as a pozzolan (Gilliland and Moran, 1949). Since this time there have been a number of ASR laboratory studies using various natural pozzolans (ACI 232, 2012), and the results support the general trend shown in Figure 9.34, that is, that the amount required depends on the composition of the

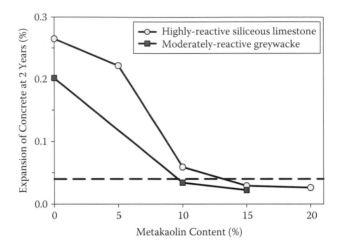

Figure 9.36 Effect of high-reactivity metakaolin on the expansion of concrete with reactive aggregate. (Data from Ramlochan, T., and Thomas, M.D.A., in *Proceedings of the 4th ACI/CANMET International Conference on the Durability of Concrete*, ed. V.M. Malhotra, ACI SP-192, Vol. I, pp. 239–251, 2000.)

pozzolan, the reactivity of the aggregate, and the amount of alkali available in the system. Unfortunately, few studies have been conducted on natural pozzolans using concrete expansion tests. The data from a study using the CPT (ASTM C 1293), two types of reactive aggregate, and a high-reactivity metakaolin from Georgia are shown in Figure 9.36. The metakaolin is very efficient in controlling ASR, and the level required to control expansion to less than 0.040% at 2 years varies from 10% for a moderately reactive aggregate to 15% for a highly reactive aggregate.

The composition of commercially available fly ashes in North America varies widely. The efficacy of fly ash in terms of controlling ASR expansion is dependent on its composition, and Class F fly ashes with low calcium contents are more effective than Class C fly ashes with high-calcium fly ashes. This is shown in Figure 9.37 with the 2-year expansion of concrete mixes containing a highly reactive siliceous limestone and 25% fly ash from various sources plotted against the calcium oxide content of the fly ash. Most fly ashes with less than 20% calcium (as CaO) were effective in controlling expansion with this aggregate, whereas the expansion for fly ashes with calcium contents greater than 20% CaO showed a trend of increasing expansion with increasing calcium content. Note that fly ashes with high alkali contents did not fit this general trend. Class C fly ashes with high calcium contents can control expansion in the CPT provided they are used at a high enough replacement level. Figure 9.38 shows that replacement levels of 50 to 60% may be required with some fly ashes.

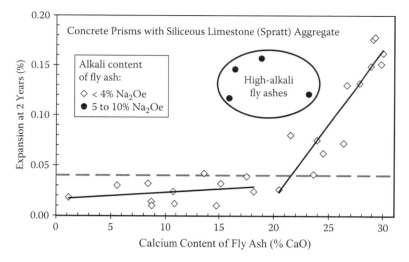

Figure 9.37 Effect of fly ash composition on the expansion of concrete with 25% fly ash and highly reactive spratt aggregates.

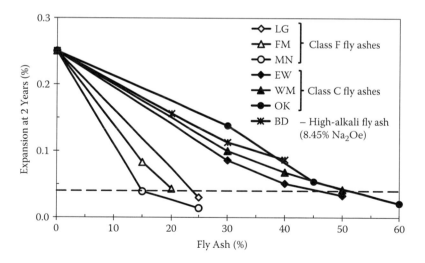

Figure 9.38 Effect of fly ash composition on the level of replacement required to suppress ASR expansion. (From Shehata, M.H., and Thomas, M.D.A., *Cement and Concrete Research*, 30(7), 1063–1072, 2000. Printed with permission from Elsevier).

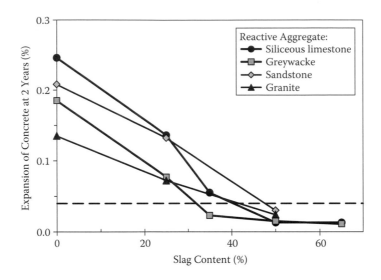

Figure 9.39 Effect of slag on the expansion of concrete with various reactive aggregates. (From Thomas, M.D.A., and Innis, F.A., *ACI Materials Journal*, 95(6), 1998. Printed with permission from the American Concrete Institute.)

The composition of slag from North American sources does not vary widely, and hence there isn't a significant variation in their performance with regards to ASR. Figure 9.39 shows the 2-year expansion of concrete prisms plotted against the slag content for four different reactive aggregates. The slag was effective in controlling expansion with all four aggregates, but the amount required ranged from 35 to 50%, depending on the aggregate type.

Silica fume is highly effective in controlling ASR expansion, and the test data available indicate that the amount of silica fume required is strongly influenced by the amount of alkali available in the concrete (Thomas and Bleszynski, 2001). In the standard CPT (ASTM C 1293), where the alkalis of the portland cement component of the mix are raised to 1.25% Na_2Oe by the addition of NaOH to the mix water, between 7 and 12% silica fume is sufficient to control expansion in the concrete prism test, depending on the reactivity of the aggregate (see Figure 9.40).

Ternary blended cements containing moderate amounts of silica fume (4 to 6%) combined with either slag or fly ash (with low or high calcium contents) have been found to be highly effective in controlling ASR expansion, as shown in Figures 9.41 and 9.42.

SCMs reduce the risk of damaging ASR by reducing the concentration of alkalis in the concrete pore solution. The pore solution changes that occur when SCMs are used were discussed in Chapter 3. The glassy silica present in SCMs reacts with the hydroxyl ions in the pore solution and calcium hydroxide to produce C-S-H with a lower Ca/Si ratio and higher alkali

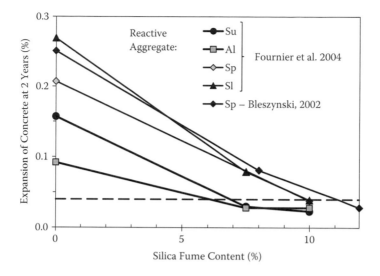

Figure 9.40 Effect of silica fume on expansion of concrete with various reactive aggregates. (Data from Fournier, B., et al., in *Proceedings of the Twelfth International Conference Alkali-Aggregate Reaction in Concrete*, ed. T. Mingshu and D. Min, Vol. I, International Academic Publishers/World Publishing Corp., Beijing, 2004, pp. 528–537.)

content than the C-S-H produced in portland cement concrete. Essentially, there is a competition between the silica in the SCM and the reactive silica in the aggregate to react with the alkali hydroxides in the system. Because SCMs are finely divided and have a high surface area, they react more rapidly than the aggregates, and provided a sufficient quantity of SCM is used, they will consume the available alkalis before deleterious reaction with the aggregate can occur. As the availability of alkalis or the reactivity of the aggregate increases, more SCM is required to prevent damaging expansion.

9.5 SULFATE ATTACK

Portland cement concrete is vulnerable to chemical attack when exposed to sulfates from either naturally occurring sources in soils or groundwaters, or industrial or agricultural environments. In North America problems due to sulfate attack have occurred in the northern Great Plains, the western United States, and the Canadian prairies (ACI 201, 2008). The hydration products of tricalcium aluminate ($3CaO.Al_2O_3$ or C_3A) are the phases of portland cement concrete that are most susceptible to sulfate attack, and sulfate-resisting portland cements are produced by limiting the amount of C_3A. Other cement hydration products, including the calcium-silicate-hydrates (C-S-H), can also

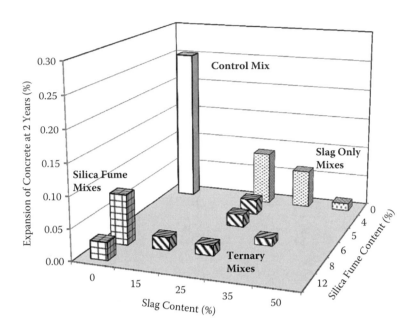

Figure 9.41 Controlling ASR expansion with ternary cement blends containing silica fume and slag. (Plotted using data from Bleszynski, R., The Performance and Durability of Concrete with Ternary Blends of Silica Fume and Blastfurnace Slag, PhD thesis, University of Toronto, 2002.)

be attacked in more severe sulfate environments. Detailed accounts of sulfate attack can be found in Thomas and Skalny (2006) and Skalny et al. (2002).

Commonly occurring reactions between sulfates and the hydration products of portland cement are shown in Figure 9.43. Calcium sulfate commonly occurring as gypsum ($CaSO_4 \cdot 2H_2O$) has a relatively low solubility (2000 mg/L water) and only attacks the aluminate-bearing phases of the cement paste, and is thus considered to be the least aggressive of the naturally occurring sulfate compounds. The reaction of calcium sulfate with the calcium aluminate hydrates and calcium monosulfoaluminate produces ettringite and generally results in expansion of the concrete.

Alkali sulfates are much more soluble than gypsum and can be present at high concentrations in groundwaters. The increased solubility, coupled with the fact that alkali sulfates attack a wider range of hydration products, means that they are considered to be more aggressive than calcium sulfate. Alkali sulfate will attack calcium hydroxide to form gypsum, and the gypsum in turn will combine with monosulfate to form ettringite. Alkali sulfates also attack the calcium aluminate hydrates, and with prolonged exposure, they may eventually attack C-S-H. Attack on the C-S-H produces gypsum and silica gel, resulting in a loss of cohesion and softening

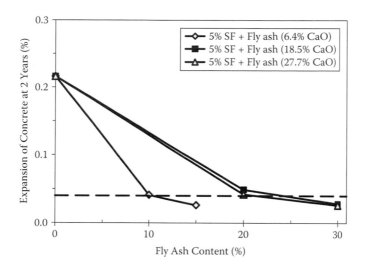

Figure 9.42 Controlling ASR expansion with ternary cement blends containing silica fume and fly ash. (From Shehata, M.H., and Thomas, M.D.A., *Cement and Concrete Research*, 32(3), 341–349, 2002. Printed with permission from Elsevier.)

of the cement paste. This may or may not be accompanied by expansion. One of the products of alkali sulfate attack is alkali hydroxide, which raises the pH of the concrete pore solution, and this stifles further reaction if the NaOH accumulates. Continued attack by alkali sulfates can only occur if the conditions promote the removal of the NaOH (or KOH) and replenish the source of Na_2SO_4 (or K_2SO_4).

Magnesium sulfate attacks the cement hydration products in a similar manner to alkali sulfates, but it is generally considered to be more aggressive because of the very low solubility of $Mg(OH)_2$ (brucite) compared with alkali hydroxide, which allows the reactions to proceed to completion provided sufficient sulfate is available. Magnesium silicate hydrate may also form as a result of the decomposition of the C-S-H. In prolonged exposure to $MgSO_4$, ettringite will eventually decompose to form gypsum and hydrated alumina (Lea, 1971). The formation of brucite can, under some circumstances, form a protective insoluble layer on the hydrates that hinders further penetration and reaction of the magnesium sulfate (Thomas and Skalny, 2006).

Under certain exposure conditions (cool, wet environments) and when a source of carbonate (CO_3^{2-}) ions is present, attack on the C-S-H by alkali and magnesium sulfates may also produce the mineral thaumasite; this is referred to in the literature as the thaumasite form of sulfate attack (TSA). The conversion of C-S-H to thaumasite also results in the loss of cohesion of the cement paste, and concrete in an advanced state of TSA completely loses integrity and turns to mush.

Normal formation of ettringite and monosulfate during first few days of hydration of portland cement

$$C_3A \quad + \quad 26H \quad + \quad 3C\bar{S}H_2 \quad \rightarrow \quad C_3A\cdot3C\bar{S}\cdot H_{32}$$

Tricalcium aluminate + Water + Gypsum → Calcium sulphoaluminate (Ettringite)

Eqn. 9.11

$$2C_3A \quad + \quad 4H \quad + \quad C_3A\cdot3C\bar{S}\cdot H_{32} \quad \rightarrow \quad 3C_3A\cdot C\bar{S}\cdot H_{12}$$

Tricalcium aluminate + Water + Ettringite → Calcium monosulphoaluminate

Eqn. 9.12

Attack by external source of calcium sulfate converting monosulfate and C-A-H to ettringite

$$C_3A\cdot C\bar{S}\cdot H_{12} \quad + \quad 16H \quad + \quad 2C\bar{S}H_2 \quad \rightarrow \quad C_3A\cdot3C\bar{S}\cdot H_{32}$$

Calcium monosulphoaluminate + Water + Gypsum → Ettringite

Eqn. 9.13

$$C_3AH_6 \quad + \quad 20H \quad + \quad 3CSH_2 \quad \rightarrow \quad C_3A\cdot3C\bar{S}\cdot H_{32}$$

Tri-calcium aluminate hydrate + Water + Gypsum → Ettringite

Eqn. 9.14

Attack by alkali sulfate on CH, C-A-H, and C-S-H (note potassium, K, behaves the same as sodium, Na)

$$CH \quad + \quad N_2\bar{S}H_{10} \quad \rightarrow \quad C\bar{S}H_2 \quad + \quad 2NH \quad + \quad 7H$$

Calcium hydroxide + Sodium sulfate → Gypsum + Sodium hydroxide + Water

Eqn. 9.15

$$2C_3AH_6 \quad + \quad 3N_2\bar{S}H_{10} \rightarrow C_3A\cdot3C\bar{S}\cdot H_{32} + \quad 2AH_3 \quad + \quad 6NH \quad + \quad 5H$$

Tri-calcium aluminate hydrate + Sodium sulfate → Ettringite + Aluminum hydroxide + Sodium hydroxide + Water

Eqn. 9.16

$$C_3S_2H_8 \quad + \quad 3N_2\bar{S}H_{10} \rightarrow \quad 3C\bar{S}H_2 \quad + 2SiO_2\cdot aq + \quad 6NH \quad + \quad 20H$$

Calcium-silicate hydrate + Sodium sulfate → Gypsum + Silica gel + Sodium hydroxide + Water

Eqn. 9.17

Attack by magnesium sulfate on C-S-H (note that magnesium sulfate also attacks CH, monosulfate, and other calcium aluminate phases in a similar manner to alkali sulfates)

$$C_3S_2H_8 \quad + \quad 3M\bar{S}H_7 \quad \rightarrow \quad 3C\bar{S}H_2 \quad + 2SiO_2\cdot aq + \quad 3MH_2 \quad + \quad 17H$$

Calcium-silicate hydrate + Magnesium sulfate → Gypsum + Silica gel + Magnesium hydroxide + Water

Eqn. 9.18

Figure 9.43 Common reactions occurring when external sulfates attack portland cement hydrates.

The major factors governing the sulfate reactions are (Thomas and Skalny, 2006):

- The type and concentration of the sulfates present in the soil or groundwater
- The nature of the exposure conditions (fluctuating temperature and moisture conditions increase the severity of the attack)

Color Figure 9.2 Cracking, spalling, and delamination of concrete cover due to corrosion of embedded reinforcement.

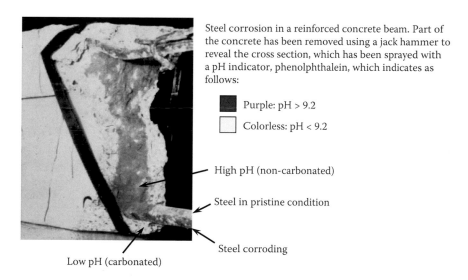

Steel corrosion in a reinforced concrete beam. Part of the concrete has been removed using a jack hammer to reveal the cross section, which has been sprayed with a pH indicator, phenolphthalein, which indicates as follows:

■ Purple: pH > 9.2
□ Colorless: pH < 9.2

High pH (non-carbonated)

Steel in pristine condition

Low pH (carbonated)

Steel corroding

Color Figure 9.12 Carbonation-induced steel corrosion in a reinforced concrete beam.

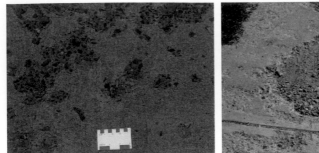

Color Figure 9.19 Examples of de-icer salt scaling of concrete driveway (left) and sidewalk (right).

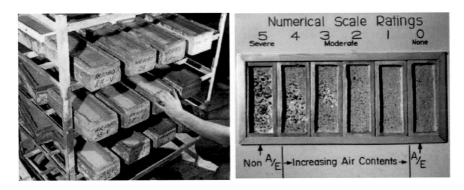

Color Figure 9.20 Laboratory salt scaling tests showing concrete slabs with dyke to contain salt solution (left) and visual rating system (right). (From Caldarone, M.A., et al., *Guide Specification for High-Performance Concrete for Bridges*, EB233, 1st ed., Portland Cement Association, Skokie, IL, 2005; Kosmatka, S.H., and Wilson, M.L., *Design and Control of Concrete Mixtures*, EB001, 15th ed., Portland Cement Association, Skokie, IL, 2011. Printed with permission from the Portland Cement Association.)

Color Figure 9.23 The author examines a slipformed concrete pavement with 50% Class C fly ash after exposure to de-icing salts over six winters.

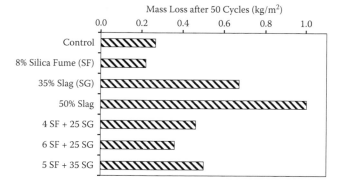

Color Figure 9.24 Construction and performance of paving slabs with combinations of portland cement (PC), silica fume (SF), and slag (SG). Note excellent field performance of concrete with 5% silica fume and 35% slag after nine winters (photo top right), but extremely poor performance in the laboratory salt scaling test (bottom).

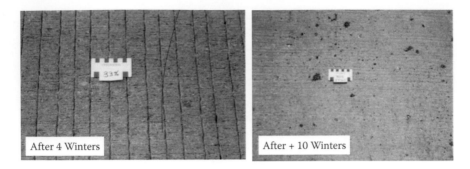

After 4 Winters

After + 10 Winters

Color Figure 9.26 Excellent performance of slipformed concrete pavement with 33% Class C fly ash on Highway 52 in Minnesota (left) and with 70% Class C fly ash on a Haul Road at Pleasant Prairie Generating Station in Wisconsin (right).

After 10 Winters

After 5 Winters

Color Figure 9.27 Satisfactory performance (moderate scaling) in slipformed pavement with 70% Class C fly ash (left) and 50% Class F fly ash (right) at Pleasant Prairie Generating Station in Wisconsin.

Color Figure 9.28 Adjacent concrete slabs on paved parking lot in Wisconsin; both slabs contain 40% Class F fly ash. Slab to left placed and finished by slipform paving machine. Slab to right placed and finished by hand.

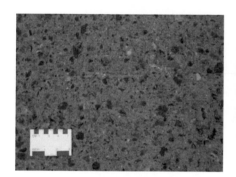 

Color Figure 9.29 High-volume fly ash concrete (left) and portland cement concrete (right) in Halifax after 12 years with approximately 100 freeze-thaw cycles per annum.

Color Figure 9.31 Cracking of concrete due to ASR in (a) bridge abutment, (b) hydraulic dam, (c) retaining wall, (d) pavement, (e) bridge piers and beams, and (f) curb and gutter.

Reaction between the alkalis (Na, K & OH) from the cement react and unstable silica, SiO$_2$, in some types of aggregate

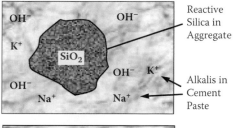

Reactive Silica in Aggregate

Alkalis in Cement Paste

The reaction produces an alkali-silica gel

The gel absorbs water from the surrounding paste …

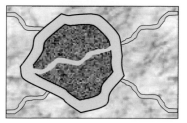

Alkali-Silica Gel

… and expands.

The internal expansion eventually leads to cracking of the surrounding concrete.

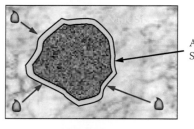

Color Figure 9.32 Mechanisms of ASR.

Cement Paste

Reactive Aggregate

Reaction Product

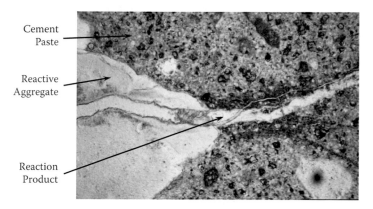

Color Figure 9.33 Thin section showing site of expansive reaction in ASR-affected concrete (width of view approximately 2-mm).

Color Figure 9.44 Photographs of 100-mm (4-in.) cubes after 5 years exposure to MgSO$_4$ solution (1.5% SO$_3$). Concrete (W/CM = 0.43 to 0.46) produced with ordinary portland cement (left), sulfate-resistant portland cement (center), and a blend of ordinary portland cement plus 20% fly ash. (Photographs copyright © BRE (UK). Reproduced with permission.)

Color Figure 9.47 Cracking of a cast-in-place reinforced bridge column due to heat-induced delayed ettringite formation and ettringite-filled gaps forming around some aggregate particles.

Figure 9.44 **(See color insert.)** Photographs of 100-mm (4-in.) cubes after 5 years exposure to $MgSO_4$ solution (1.5% SO_3). Concrete (W/CM = 0.43 to 0.46) produced with ordinary portland cement (left), sulfate-resistant portland cement (center), and a blend of ordinary portland cement plus 20% fly ash. (Photographs copyright © BRE (UK). Reproduced with permission.)

- The permeability of the concrete, which affects the rate at which the destructive SO_4^{2-} ions can penetrate the concrete and attack the hydrates
- The composition of the hydrated cement phases in the concrete

Portland cement concrete is made resistant to sulfate attack by using Type V sulfate-resistant cement that is low in C_3A and by ensuring that the concrete has a low W/CM. In extremely severe sulfate environments it may be necessary to protect the concrete from direct contact by using a suitable barrier system. It has long been recognized that some pozzolans and slag can increase the resistance of concrete to sulfate attack, and their use offers an alternative to using sulfate-resisting portland cement. Figure 9.44 shows photographs of concrete cubes exposed to sulfate solution and the improved performance of concrete that can be achieved by using either Type V sulfate-resisting portland cement or a sufficient amount of low-calcium fly ash. However, not all SCMs are suitable for this purpose, and the effectiveness of a particular SCM will depend on its composition, the level of replacement used, and the composition of the portland cement (e.g., C_3A content) with which the SCM is combined. As with using sulfate-resisting portland cement, when SCMs are used in concrete exposed to sulfates, it is important to ensure that a low W/CM is maintained.

Most specifications control sulfate attack by (i) characterizing the severity of the exposure condition and (ii) selecting suitable protective measures. An example of this approach is the recently revised ACI 318 Building Code. Table 9.8 shows the sulfate exposure classes of ACI 318. There are no special requirements for concrete in Class S0, but for the other classes there are requirements in terms of minimum strength, maximum W/CM, and the type or performance of the cementing materials. Basically any combination of portland cement, blended cement, and SCMs can be used provided they meet certain performance requirements when tested in ASTM C 1012:

Table 9.8 Sulfate exposure classes

Severity	Class	Condition	
		Water-soluble sulfate in soil (% SO₄ by mass)	Dissolved sulfate in water (ppm SO₄)
Not applicable	S0	<0.10	<150
Moderate	S1	≥0.10, <0.20	≥150, <1500
Severe	S2	≥0.20, ≤2.00	≥1500, ≤10,000
Very severe	S3	>2.00	>10,000

Source: Modified from ACI 318-08. Produced with permission from the American Concrete Institute.

Standard Test Method for Length Change of Hydraulic-Cement Mortars Exposed to a Sulfate Solution. In this test mortar bars are immersed in a 5% Na_2SO_4 solution and the length change is monitored for a period up to 18 months. Cements to be used in the different sulfate exposure classes must meet the expansion limits specified in Table 9.9.

Figure 9.45 shows expansion data from ASTM C 1012 testing from the author's laboratory and from Ramlochan and Thomas (2000). The expansion behavior of the mortars produced with portland cement show expected behavior, with Type I cement expanding the most, Type II meeting the requirement for moderate resistance (<0.10% at 6 months), and Type V meeting the requirement for high resistance (<0.10% at 12 months). Type V cement does not have sufficient sulfate resistance to be used in Class 3 sulfate exposure according to the requirements of ACI 318 (see Table 9.9), but may achieve a sufficient level of resistance if combined with a suitable SCM.

The performance of blended cements is dependent on the amount and type of pozzolan or slag used, as shown in Figure 9.43, and some blends have superior performance to Type V sulfate-resistant portland cement.

A blend of Type I (high-C_3A) portland cement with slag performs well, and blends with 25 and 35% slag meet the requirements for moderate and high sulfate resistance, respectively, and the blend with 50% meets the ACI

Table 9.9 Requirements for blends of portland cement, pozzolans, and slag to be used in sulfate exposure classes

Sulfate exposure class	Maximum expansion in ASTM C 1012		
	At 6 months	At 12 months	At 18 months
S1: Moderate	0.10%		
S2: Severe	0.05%	0.10%ᵃ	
S3: Very severe			0.10%

Source: ACI 318-08. Produced with permission from the American Concrete Institute.

ᵃ The 12-month limit for S2 applies only when the measured expansion exceeds the 6-month maximum expansion limit.

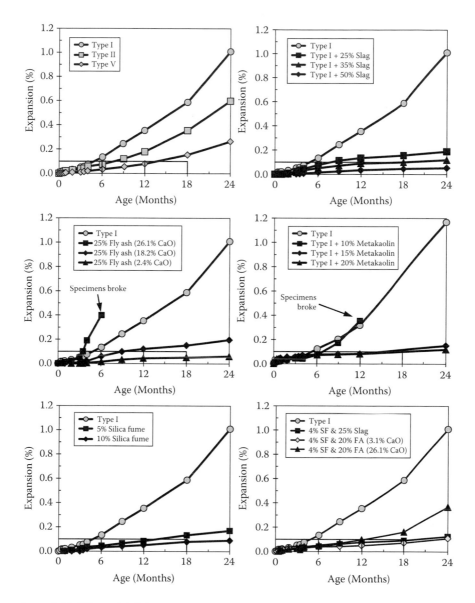

Figure 9.45 Effect of SCM on expansion of mortars stored in 5% Na$_2$SO$_4$ solution as per ASTM C1012. (Data for metakaolin from Ramlochan, T., and Thomas, M.D.A., in *Proceedings of the 4th ACI/CANMET International Conference on the Durability of Concrete*, ed. V.M. Malhotra, ACI SP-192, Vol. 1, pp. 239–251, 2000.)

318 requirement to be used in Class S3 sulfate exposure. However, it should be noted that the sulfate resistance of mortars or concrete containing slag decreases as the alumina content of the slag increases (Hooton and Emery, 1990). Although slags from North American sources tend to have relatively low alumina contents ($<10\%$ Al_2O_3), imported slags may have higher contents and exhibit inferior sulfate resistance.

Combinations of high-C_3A portland cement and 25% low-calcium fly ash also meet the ACI 318 requirements for the S3 exposure class. However, the performance of fly ash cements in sulfate exposure is strongly dependent on the composition of the fly ash. Low-calcium fly ashes generally perform well, but the behavior of fly ashes with calcium contents above 15% CaO is variable, and high-calcium fly ashes with more than 20% CaO usually perform poorly and often decrease the sulfate resistance compared to the control with 100% Type I cement (Thomas et al., 1999). The good performance of Class F fly ash in sulfate resistance tests and inferior performance of high-calcium fly ash have been confirmed by a number of studies (Dunstan, 1980; Mehta, 1986; von Fay and Pierce, 1989; Tikalsky and Carrasquillo, 1992).

The performance of blended cement containing metakaolin, a natural pozzolan produced by thermal activation of relatively pure kaolin clay, depends strongly on the level of pozzolan present (Ramlochan and Thomas, 2000). Low levels of metakaolin (e.g., 10%) tend to reduce the sulfate resistance, but at higher levels, the resistance is much improved and blended cement containing 20% metakaolin appears to meet the ACI 318 requirements for sulfate exposure Class S3.

Silica fume is effective in improving sulfate resistance in the ASTM C 1012 test, and a high level of resistance ($<0.10\%$ at 12 months) suitable for Class S2 exposure can be achieved with just 5% silica fume. A blended cement with 10% silica fume meets the requirements of ACI 318 for Class S3 exposure. Ternary blends with silica fume and either fly ash or slag also offer a high level of resistance, although when a high-calcium fly ash is used, there is some long-term expansion.

The beneficial effect of SCMs on the sulfate resistance of mortar and concrete may be ascribed to a number of mechanisms, including:

- Reduced permeability
- Dilution of the C_3A phases and CH (both participants in reactions with sulfates) as a result of the partial replacement of portland cement
- Consumption of CH by pozzolanic reaction
- Alteration of hydrated aluminate phases, making them more resistant; for example, the presence of reactive silica may favor the formation of strätlingite (C-A-S-H)

The reduction in sulfate resistance observed with high-calcium fly ashes compared with other pozzolans may be attributed to some or all of the following:

- Reduced pozzolanity and hence CH consumption
- Presence of crystalline phases that may participate in sulfate reactions, such as C_3A and anhydrite
- Presence of reactive aluminates in glass phase
- Production of reactive calcium aluminates (e.g., gehlenite)

9.6 HEAT-INDUCED DELAYED ETTRINGITE FORMATION

Under certain circumstances, concrete that is exposed to elevated temperatures at early ages, either deliberately due to the application of heat to accelerate early strengths or adventitiously due to autogenous temperature rise, may experience delayed ettringite formation (DEF) and suffer expansion and cracking. A schematic of the steps in the process of damaging DEF is shown in Figure 9.46. At elevated temperatures, the ettringite that normally forms at early ages in concrete dissolves incongruently, and much of the sulfate and alumina becomes trapped in the rapidly forming C-S-H. When the concrete is subsequently exposed to normal temperatures and moisture in service, the sulfate and alumina are slowly released by the C-S-H to form ettringite. This delayed ettringite formation is sometimes, but not always, accompanied by expansion. The risk of damage due to DEF increases as the peak temperature during curing increases above 70°C (158°F) and as the C_3A, SO_3, C_3S, Na_2Oe, and fineness of the cement increase. Damage due to DEF does not occur unless the concrete is exposed to elevated temperature. Damage due to DEF is not a widespread problem, and there have been very few cases where DEF has been unequivocally established as the sole cause of deterioration in a structure.

Figure 9.47 shows a cast-in-place reinforced concrete bridge column that has suffered cracking that is solely attributed to DEF (Thomas et al., 2008). The concrete was not heat cured, but it is a massive element that was cast at high ambient temperatures and using a concrete with high cement content, the cement being of very high fineness. It was estimated (Thomas et al., 2008) that the maximum internal temperature of the concrete would have certainly exceeded 70°C (58°F) and possibly even 80°C (176°F). An electron micrograph of a polished section of concrete taken from this column is also shown in Figure 9.47. DEF causes the paste to expand, and since the aggregates are volumetrically stable, gaps open up around some aggregates; this is considered to be a reliably diagnostic feature of DEF expansion. Some of the gaps eventually become filled with ettringite.

T ~ 20°C (70°F) with no heat curing

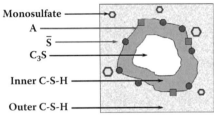

Monosulfate

Ettringite

C_3S

Inner C-S-H

Outer C-S-H

When cured at normal ambient temperature, ettringite and monosulfate are formed as part of the outer hydration products and there is no expansion as there is plenty of space

T > 70°C (160°F)

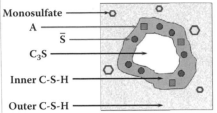

Monosulfate

A

$\bar{S}$

C_3S

Inner C-S-H

Outer C-S-H

When cured at elevated temperatures, the incongruent dissolution of ettringite results in both sulfate and alumina being encapsulated in the rapidly forming inner C-S-H

T drops from > 70°C (160°F) to 20°C (70°F)

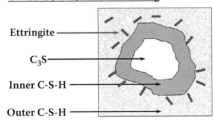

Monosulfate

A

$\bar{S}$

C_3S

Inner C-S-H

Outer C-S-H

After exposure to elevated temperature and during subsequent exposure to moisture at normal ambient temperature, sulfate and alumina slowly released from the inner C-S-H

T ~ 20°C (70°F) after heat curing

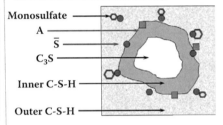

Monosulfate

A

$\bar{S}$

C_3S

Inner C-S-H

Outer C-S-H

... and reacts with monosulfate (or calcium hydroxide) ...

T ~ 20°C (70°F) after heat curing

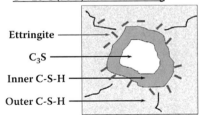

Ettringite

C_3S

Inner C-S-H

Outer C-S-H

... to form ettringite in the fine pores of the outer C-S-H

T ~ 20°C (70°F) after heat curing

Ettringite

C_3S

Inner C-S-H

Outer C-S-H

There is no longer space to accommodate the ettringite and its delayed formation leads to expansion of the paste (under some circumstances)

Figure 9.46 Schematic showing steps in delayed ettringite formation.

Figure 9.47 **(See color insert.)** Cracking of a cast-in-place reinforced bridge column due to heat-induced delayed ettringite formation and ettringite-filled gaps forming around some aggregate particles.

The most obvious precaution for minimizing the risk of DEF damage is to ensure that maximum internal concrete temperature does not exceed 60 to 70°C (140 to 158°F). If this is not possible or an additional precautionary measure is required, then the risk can be further minimized by selecting a portland cement that is not prone to heat-induced expansion or by using a sufficient quantity of an appropriate SCM.

One of the first published laboratory studies on DEF included data that demonstrated the ability of fly ash and slag to control expansion in heat-cured mortars (Ghorab et al., 1980). Since then there have been a number of studies showing the impact of pozzolans and slag. Figure 9.48 shows the expansion of mortar bars that were initially heat cured at 95°C (203°F) and subsequently stored in limewater (Ramlochan et al., 2004). SCMs that contain significant quantities of alumina, such as fly ash, slag, and metakaolin, were found to be effective in suppressing expansion when used at common replacement levels (25% fly ash or slag and 8% metakaolin). Silica fume, which contains a negligible amount of alumina, was not effective in controlling expansion at a replacement level of 8% and had to be used at levels above 15% to inhibit expansions up to ages of 1500 days (Ramlochan et al., 2003). Ramlochan et al. (2003, 2004) concluded that the available alumina in the SCM is responsible for the beneficial effect on DEF expansion, but the precise mechanism is not understood.

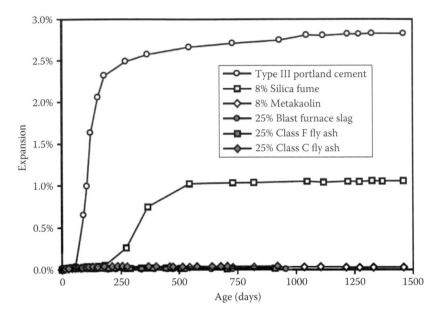

Figure 9.48 Effect of SCM on the expansion of heat-cured mortars stored in limewater. (From Ramlochan, T., et al., *Cement and Concrete Research*, 34, 1341–1356, 2004. With permission.)

9.7 PHYSICAL SALT ATTACK

Physical salt attack or physical sulfate attack is differentiated from chemical attack in that there is no chemical alteration of the cement hydration products. However, it has been argued (Skalny et al., 2002) that the resulting deterioration is due to a chemical process, such as the crystallization of salts from solution or the cyclic hydration-dehydration of salts, and thus the term *physical attack* is not appropriate.

Regardless of terminology, this type of attack commonly occurs with sodium sulfate, but may also occur with other nonsulfate salts, such as sodium carbonate (Haynes et al., 1996). This type of attack is mainly restricted to relatively low-quality concrete (high W/CM) that is exposed to water containing dissolved salts on one face and a relatively dry environment on another. Examples would be residential slabs or foundation walls, with inadequate drainage and waterproofing. The salt solution is "wicked" up into the porous concrete by capillary action and migrates to the drying face where the salts are concentrated as the water evaporates and eventually crystallize from solution in the pores close to the surface. Oftentimes the consequence is merely an unsightly but nondeleterious efflorescence, with salt crystals growing on the surface. However, in some cases the salt crystals grow below the surface, subflorescence, and can cause disruption

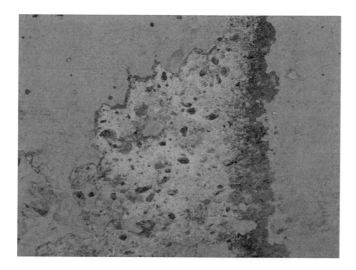

Figure 9.49 Crystallization or hydration-dehydration of salts leading to spalling of concrete surface.

and spalling of the surface layers of concrete (Figure 9.49). The damage is exacerbated by fluctuations in temperature and humidity at the surface, which can lead to cyclic hydration and dehydration of the salt. For example, changes in temperature and humidity can result in the following transformation of sodium sulfate (Folliard and Sandberg, 1994):

$$Na_2SO_4 \xleftrightarrow{\;H_2O\;} Na_2SO_4 \cdot 10H_2O \tag{9.11}$$

Thenardite $\leftrightarrow$ Mirabilite

The decahydrate, mirabilite, has a much higher molar volume than anhydrous thenardite, and hydration leads to a significant increase in solid volume. The repeated cycles of volume change that accompany cyclic hydration-dehydration are analogous to the volume changes that occur with water-ice transformations during freeze-thaw cycles. Indeed, the ASTM C 88 aggregate soundness test takes advantage of the volume changes that occur during wetting and drying cycles with sodium or magnesium sulfate to evaluate the freeze-thaw resistance of aggregates.

It is universally accepted that the most effective methods for preventing physical salt attack of concrete are to ensure proper building practices (e.g., use of vapor barriers below slabs or waterproofing membranes on foundation walls, and provision of an adequate drainage layer to ensure capillary discontinuity between the groundwater and the concrete) to prevent continuous wicking of moisture into the concrete and to ensure that the concrete is sufficiently resistant to moisture flow; W/CM ≤ 0.45 has been recommended (Haynes et al., 1996).

The influence of SCMs on physical salt attack is not clear. Haynes et al. (1996) recommended the use of an SCM, such as fly ash or, in severe cases, silica fume, to further reduce the permeability of low-W/CM concrete. Such advice is not consistent with the conclusions from PCA's long-term test program at the sulfate exposure test facility in Sacramento, California, which found the main mechanism of deterioration to be physical salt attack, and that SCMs did not generally improve performance, and some SCMs (silica fume and slag) may actually reduce resistance to attack (Stark, 2002). Irassar et al. (1996) also found that high levels of SCM increased the surface scaling of concrete half-buried in soil contaminated with sodium sulfate, but improved the performance of buried concrete that was permanently saturated with solution. They attributed the physical damage above grade to the high capillary suction of concrete with SCM.

Further research is clearly needed to elucidate the mechanistic aspects of physical salt attack and the role played by SCMs.

9.8 OTHER FORMS OF CHEMICAL ATTACK

A number of chemical compounds, particularly acids, can attack cement hydration products, leading to deterioration of even high-quality concrete. A list of some of the chemicals that are known to attack concrete is presented in Table 9.10.

Acid attack on concrete results in the decalcification of the CH and C-S-H in the paste fraction, and can eventually lead to softening and disintegration

Table 9.10 Chemical compounds that attack concrete

Promote rapid deterioration	Promote moderate deterioration	
Aluminum chloride	Ammonium salts	Potassium hydroxide (>25%)
Calcium bisulfite	Castor oil	Rapeseed oil
Hydrochloric acid	Cocoa bean oil	Slaughterhouse waste
Hydrofluoric acid	Cocoa products	Sodium bisulfate
Nitric acid	Coconut oil	Sodium bisulfite
Sulfuric acid	Cottonseed oil	Sodium hydroxide (>20%)
Sulfurous acid	Fish liquor	Sulfite liquor
	Mustard oil	Tanning liquor
	Perchloric acid	Zinc refining solutions
	Potassium dichromate	

Source: From Portland Cement Association (PCA), *Types and Causes of Concrete Deterioration*, PCA Report IS536, Portland Cement Association, Skokie, IL, 2002. Produced with permission from the Portland Cement Association.

of the binder phase. If calcareous aggregates (limestone and dolostone) are present, they may also be attacked. The rate of attack increases as the pH decreases, but is also dependent on the solubility of the salt that is formed and conditions of exposure (e.g., rate at which the acid is replenished and the reaction products removed). Attack by acids that form soluble compounds, such as hydrochloric acid, which forms calcium chloride according to Equation 9.12, proceeds rapidly because the reaction products are readily leached away. Acids, such as phosphoric acid, are not as aggressive because the reaction product, calcium phosphate, is not readily removed from the surface and may protect unreacted material from further attack.

$$Ca(OH)_2 + 2HCl \rightarrow CaCl_2 + 2H_2O \qquad (9.12)$$

The rate of attack is generally negligible at pH > 6.5, but the rate of attack increases as the pH reduces below 6.5. The rate of attack at pH <6.5 can be reduced by proper selection of materials (e.g., use of SCM) and proportioning (low W/CM). No portland cement concrete, regardless of its composition, can withstand exposure to water with pH < 3.0, and in such conditions a protective barrier or treatment must be used (ACI 201.2R).

Ammonium salts, such as ammonium nitrate (NH_4NO_3) and ammonium chloride (NH_4Cl), may produce moderate to rapid chemical attack because in the high pH of concrete the ammonium ions are converted to ammonia (NH_3) and hydrogen (H_2) gas, and are replaced by calcium ions from the CH and C-S-H. Chlorides and nitrates of magnesium, aluminum, and iron can all cause distress, but are not as aggressive as ammonium salts. Strong alkaline solutions with concentrations in excess of 20% can also cause concrete disintegration.

The rate of chemical attack by acids and other compounds can be reduced by the proper section of materials and proportioning concrete to have a low W/CM to reduce permeability. The use of SCMs can further reduce permeability and increase the resistance of the concrete to aggressive environments. However, in severe environments, concrete has to be protected by surface treatments. PCA IS001 provides guidance on protecting concrete from deleterious substances, including acids, salts and alkalies, petroleum oils, coal tar distillates, solvents and alcohols, vegetable oils, animal fats and fatty acids, and a range of other miscellaneous products.

9.9 ABRASION, EROSION, AND CAVITATION

Concrete floors and pavements exposed to heavy vehicular traffic are subjected to abrasion, and hydraulic and marine structures may be exposed to abrasion, erosion, and cavitation. The abrasion resistance of concrete is dependent on the strength of the concrete and the type of aggregate.

Concrete produced with the same aggregate and proportioned to have the same compressive strength can be expected to have the same abrasion resistance, regardless of the presence of SCMs. Silica fume concrete has excellent abrasion-erosion resistance, due to its high strength, and one of the first major uses of silica fume in the United States exploited this characteristic in repairs to erosion-damaged concrete in the Kinzua Dam in Pennsylvania (Holland et al., 1986).

Chapter 10

Specifications

There are many specifications worldwide for supplementary cementing materials (SCMs) and for blended cements containing SCMs. This chapter discusses American Society for Testing and Materials (ASTM) specifications only.

10.1 ASTM SPECIFICATIONS FOR SCMS

SCMs that are used as a separate addition at the concrete mixer are covered by the following ASTM specifications:

ASTM C 618   *Standard Specification for Coal Fly Ash and Raw or Calcined Natural Pozzolan for Use in Concrete*

ASTM C 989   *Standard Specification for Ground Granulated Blast-Furnace Slag for Use in Concrete and Mortars*

ASTM C 1204 *Standard Specification for Silica Fume Used in Cementitious Mixtures*

ASTM C 618 covers Class F and Class C fly ashes (see Table 2.5 for definition of classes), and Class N raw or calcined natural pozzolans. ASTM C 989 classifies slag as Grade 80, 100, or 120 based on the slag activity index (see Table 2.7).

The chemical requirements for SCMs in these specifications are summarized in Table 10.1.

The physical requirements are shown in Table 10.2. There is a strength activity requirement for all the SCMs, although the terminology used is different for each specification. The strength activity index is based on the strength of 50-mm (2-in.) mortar cubes compared with a mixture without SCM. For fly ashes and natural pozzolans, ASTM C 618 requires that a mixture made with 20% SCM achieves at least 75% of the strength of the control at 7 and 28 days. The water content of mortars with fly ash or natural pozzolan is adjusted to produce the same flow as the control mixture,

Table 10.1 Chemical requirements for SCMs

Property	ASTM C 618			C 989 slag	C 1240 silica fume
	Class N	*Class F*	*Class C*		
$SiO_2 + Al_2O_3 + Fe_2O_3$ (% minimum)	70.0	70.0	50.0		
SiO_2 (% minimum)					85.0
Sulfate, SO_3 (% maximum)	4.0	5.0	5.0	4.0	
Sulfide sulfur, S (% maximum)				2.5	
Moisture (% maximum)	3.0	3.0	3.0		3.0
LOI (% maximum)	10.0	6.0	6.0		6.0

Table 10.2 Physical requirements for SCMs

Property	ASTM C 618			C 989 slag	C 1240 silica fume
	Class N	*Class F*	*Class C*		
Fineness, retained on 45 μm sieve (% maximum)	34	34	34	20	10
Specific surface (m^2/g minimum)					15
Strength activity at 7 days (% minimum of control)	75	75	75	Table 2.7	105
Strength activity at 28 days (% minimum of control)	75	75	75		
Water requirement (% maximum of control)	115	105	105		
Autoclave expansion or contraction (% maximum)	0.8	0.8	0.8		

and there is a limit on the water requirement, as shown in Table 10.2. The slag activity index is determined using 50% slag, and the water content of the slag and control mixture is adjusted to give a certain flow. Alkali and strength limits are placed on the reference cement used to test the mortars. The requirements for strength activity depend on the grade of the slag and are reported in Table 2.7. For silica fume, the accelerated pozzolanic activity is measured at 7 days on mortars with 10% silica fume at a fixed W/CM with flow adjusted using a high-range water-reducing admixture. The mortar cubes are cured at laboratory temperature for 1 day and then at 65°C (149°F) for the remaining 6 days.

There is a fineness requirement for all SCMs and a minimum specific surface for silica fume. Fly ashes and natural pozzolans are also required to meet a soundness requirement using an autoclave expansion test; the purpose of this test is to determine whether there is potentially damaging free lime (CaO) or periclase (MgO) present.

There are optional requirements to demonstrate the effectiveness of the SCM in terms of controlling alkali-silica reaction (ASR) and sulfate attack. The requirements for fly ash, natural pozzolans, and silica fume are in the main body of ASTM C 618 and C 1240, whereas the requirements for slag are in the appendices to ASTM C 989.

The requirements for controlling ASR are based on expansion testing of mortar bars containing highly reactive Pyrex glass (ASTM C 441) and differ for the various SCMs as follows:

- For fly ash and natural pozzolans, the test mixture with the SCM should not expand more than a control mixture made with low-alkali cement (test duration is 14 days).
- For slag, either the job cement plus slag should not expand by more than 0.02% at 14 days or if the job cement is not known, the slag should reduce the 14-day expansion of a mixture with high-alkali cement by at least 75% when compared with a mix with high-alkali cement on its own.
- For silica fume, the reduction in the 14-day expansion due to the silica fume must be at least 80% when compared to a control mix with the same high-alkali cement.

The requirements for sulfate resistance for fly ash, natural pozzolans, and slag are based on expansion testing of mortar bars stored in 5% Na_2SO_4 solution (ASTM C 1012) and may be summarized as follows:

- For moderate sulfate resistance the expansion of the test mixture should be less than 0.10% at 6 months.
- For high sulfate resistance the expansion of the test mixture should be less than 0.05% at 6 months.
- For very high sulfate resistance the expansion of the test mixture should be less than 0.05% at 12 months; only ASTM C 1240 for silica fume includes this criterion.
- ASTM C 618 includes an alternative approach (Procedure B) to demonstrate the sulfate resistance of fly ash and natural pozzolans. In this case the 6-month expansion of the test mixture must not exceed that of a control mortar produced with sulfate-resistant portland cement.

10.2 ASTM SPECIFICATIONS FOR BLENDED CEMENTS—ASTM C 595

Blended cements are covered by ASTM C 595: *Standard Specification for Blended Hydraulic Cements*. The specification pertains to hydraulic cements containing pozzolans or slag blended with portland cement or portland cement clinker, or slag blended with lime. The specification was revised in 2008, and the revised version now covers two types of blended cement:

Type IS: Portland blast furnace slag cement
Type IP: Portland pozzolan cement

A suffix (X) is added to the type of cement, where X is the target level of slag or pozzolan in the cement. Further suffices may be added to indicate if the blended cement has one or more special properties as follows:

(A) Air entraining
(MS) Moderate sulfate resistance
(HS) High sulfate resistance
(MH) Moderate heat of hydration
(LH) Low heat of hydration

So, for example, a blended cement containing 50% slag with high sulfate resistance would be designated Type IS(50)(HS).

Pozzolans and slag are also permitted in ternary blended cement, which are comprised of three cementitious materials, one of which is portland cement and the other two being either (1) two different pozzolans, (2) a pozzolan and slag, (3) a pozzolan and ground limestone, or (4) slag and ground limestone. A ternary blended cement with 60% portland cement, 20% slag, and 20% pozzolan would be named Type IT(P20)(S20).

There are limited requirements imposed on the individual components of the blended cement. For example, the portland cement does not have to meet ASTM C 150, and the pozzolans or slag do not have to meet ASTM C 618, ASTM C 989, or ASTM C 1240. Fineness and strength activity requirements are imposed on pozzolans to be used in Type IP cements and on the slag to be used in Type IS cements with slag contents below 25%. The blended cement must meet both chemical and physical (including strength) requirements.

10.3 ASTM PERFORMANCE SPECIFICATION FOR HYDRAULIC CEMENTS—ASTM C 1157

Blended cements are also covered under ASTM C 1157: *Standard Performance Specification for Hydraulic Cement*. This specification only gives performance requirements; there are no restrictions on the composition of the finished cement or its constituents. There are six types of cement covered by this specification:

Type GU: Hydraulic cement for general construction
Type HE: High early strength

Type MS: Moderate sulfate resistance
Type HS: High sulfate resistance
Type MH: Moderate heat of hydration
Type LH: Low heat of hydration

There is also an additional option that may be invoked:

Option R: Low reactivity with alkali-reactive aggregates

The special properties of the different types of cement can be achieved with or without SCMs at the option of the producer.

The different hydraulic cement types covered by ASTM C 1157 must meet a number of physical requirements (including strength), but there are no chemical requirements in the specification.

The setting time, autoclave, and mortar bar expansion requirements are the same for all cement types (the latter two of these are intended to protect from unsoundness due to free lime, periclase, and excess sulfate); otherwise, the requirements vary according to the type of cement as follows:

- Type HE cement is the only cement that has a 1-day strength requirement; it also has a higher 3-day strength requirement than the other types.
- Type MS and HS cements must meet sulfate resistance requirements. Type HS cements have lower early-age strength requirements, as low-C_3A cements or blended cements that meet the requirements of Type HS often show slower strength development.
- Type MH and LH cements must meet heat of hydration requirements, but there are lower early-age strength requirements for these cements. It is not possible to produce high early strength and low heat with portland cement-based cements.

This approach allows the purchaser to choose the performance requirements required for the job and gives the cement manufacturer the flexibility to meet those requirements using the materials available.

References

Abrams, D.A. 1918. *Design of concrete mixtures*. Bulletin 1, PCA LS001. Lewis Institute, Structural Materials Research Laboratory. Skokie, IL.

ACAA. 2006. 2005. Coal combustion product (CCP) production and use survey. American Coal Ash Association. www.acaa-usa.org. Farmington Hills, MI.

ACI 116. 2000. *Cement and concrete terminology*. ACI Committee 116 Report, ACI 116R-00. American Concrete Institute, Farmington Hills, MI.

ACI 201. 2008. *Guide to durable concrete*. ACI Committee 201 Report, ACI 201.2R-08. American Concrete Institute, Farmington Hills, MI.

ACI 207. 1995. *Effect of restraint, volume change, and reinforcement on cracking of mass concrete*. ACI Committee 207 Report, ACI 207.2R-95. American Concrete Institute, Farmington Hills, MI.

ACI 232. 2003. *Use of fly ash in concrete*. ACI Committee 232 Report, ACI 232.2R-03. American Concrete Institute, Farmington Hills, MI.

ACI 232. 2012. *Use of raw or processed natural pozzolans in concrete*. ACI Committee 232 Report, ACI 232.1R-12. American Concrete Institute, Farmington Hills, MI.

ACI 233. 2003. *Slag cement in concrete and mortar*. ACI Committee 233 Report, ACI233R-03. American Concrete Institute, Farmington Hills, MI.

ACI 234. 2006. *Guide for the use of silica fume in concrete*. ACI Committee 234 Report, ACI 234R-06. American Concrete Institute, Farmington Hills, MI.

ACI 305. 2006. *Specification for Hot Weather Concreting*. ACI Committee 305 Report, ACI 305.1-06. American Concrete Institute, Farmington Hills, MI.

ACI 318. 2008. *Building Code Requirements for Structural Concrete (ACI 318-08) and Commentary*. ACI Committee 318 Report, ACI 318-08. American Concrete Institute, Farmington Hills, MI.

Aitcin, P.-C. 1998. *High-performance concrete*. E&FN Spon, London.

Aitcin, P.-C. 1999a. Demystifying autogenous shrinkage. *Concrete International*, November, pp. 54–56.

Aitcin, P.-C. 1999b. Does concrete shrink or does it swell? *Concrete International*, December, pp. 77–80.

Aitcin, P.-C., and Neville, A. 1993. High-performance concrete demystified. *Concrete International*, 15(1), 21–26.

Anon. 1914. An investigation of the pozzolanic nature of coal ashes. *Engineering News*, 71(24), 1334–1335.

Bamforth, P.B. 1980. *In situ* measurement of the effect of partial portland cement replacement using either fly ash or ground granulated blast-furnace slag on the performance of mass concrete. In *Proceedings of Institution of Civil Engineers*, London, September, Part 2, Vol. 69, pp. 777–800.

Bamforth, P.B. 1984. Heat of hydration of fly ash concrete and its effect on strength development. In *Ashtech '84 Conference*, London, pp. 287–294.

Barneyback, R.S., Jr., and Diamond, S. 1981. Expression and analysis of pore fluids from hardened cement paste and mortars. *Cement and Concrete Research*, 11, 279–285.

Barrow, R.S., Hadchiti, K.M., Carrasquillo, P.M., and Carrasquillo, R.L. 1989. Temperature rise and durability of concrete containing fly ash. In *Proceedings of the Third International Conference on the Use of Fly Ash, Silica Fume, Slag and Natural Pozzolans in Concrete* (ed. V.M. Malhotra). ACI SP-114, Vol. 2. American Concrete Institute, Detroit, pp. 331–347.

Bentur, A., and Cohen, M.D. 1987. Effect of condensed silica fume on the microstructure of the interfacial zone in portland cement mortars. *Journal of the American Ceramic Society*, 70(10), 738–743.

Bentur, A., Diamond, S., and Berke, N.S. 1997. *Steel corrosion in concrete*. E&FN Spon, London.

Bernhardt, C.J. 1952. SiO$_2$-Stov som Cementtilsetning [SiO2 dust as an admixture to cement]. *Betongen Idag*, 17(2), 29–53.

Berry, E.E., Hemmings, R.T., Langley, W.S., and Carette, G.C. 1989. Beneficiated fly ash: Hydration, microstructure, and strength development in portland cement systems. In *Proceedings of the Third International Conference on Fly Ash, Silica Fume, Slag, and Natural Pozzolans in Concrete* (ed. V.M. Malhotra). SP-114, Vol. 1. American Concrete Institute, Farmington Hills, MI, pp. 241–273.

Best, J.F., and Lane, R.O. 1980. Testing for optimum pumpability of concrete. *Concrete International*, 2(10), 9–17.

Bhatty, M.S.Y., and Greening, N.R. 1987. Some long-time studies of blended cements with emphasis on alkali-aggregate reaction. Presented at Proceedings of 7th International Conference on Alkali-Aggregate Reactions, Ottawa, 1986. Noyes Publications, Park Ridge, NJ.

Bisaillon, A., Rivest, M., and Malhotra, V.M. 1994. Performance of high-volume fly ash concrete in large experimental monoliths. *ACI Materials Journal*, 91(2), 178–187.

Bittner, J.D., and Gasiorowski, S. 1999. Five years of commercial fly ash beneficiation by Separation Technologies, Inc. In *1999 International Ash Utilization Symposium*, October 18–20, pp. 554–560.

Bleszynski, R. 2002. The performance and durability of concrete with ternary blends of silica fume and blastfurnace slag. PhD thesis, University of Toronto.

Bleszynski, R., Hooton, R.D., Thomas, M.D.A., and Rogers, C.A. 2002. Durability of ternary blend concrete with silica fume and blast-furnace slag: Laboratory and outdoor exposure site studies. *ACI Materials Journal*, 99(5), 499–508.

Bouzoubaa, N., Tamtsia, B., Zhang, M.H., Chevrier, R.L., Bilodeau, A., and Malhotra, V.M. 2006. Carbonation of concrete incorporating high volumes of fly ash. In *Proceedings of the Seventh CANMET/ACI International Conference on the Durability of Concrete* (ed. V.M. Malhotra). ACI SP-234. American Concrete Institute, Farmington Hills, MI, pp. 283–304.

Bouzoubaa, N., Zhang, M.H., Malhotra, V.M., et al. 1997. The effect of grinding on physical properties of fly ashes and portland cement clinker. *Cement and Concrete Research*, 27(12), 1861–1974.

British Cement Association (BCA). 1999. *Concrete through the ages—From 7000 BC to AD 2000*. British Cement Association, Crowthorne, Berks.

Brooks, J.J., Wainwright, P.J., and Boukendakji, M. 1992. Influence of slag type and replacement level on strength, elasticity, shrinkage, and creep of concrete. In *Proceedings of the Fourth International Conference on Fly Ash, Silica Fume, Slag, and Natural Pozzolans in Concrete*. SP-132, Vol. 2. American Concrete Institute, Farmington Hills, MI, pp. 1325–1341.

Broomfield, J.P. 1997. *Corrosion of steel in concrete*. E&FN Spon, London.

Butler, W. 1981. Super-fine fly ash in high-strength concrete. In *Concrete 2000 economic and durable construction through excellence*. Vol. 2. E&FN Spon, Glasgow, p. 1825.

Buttler, F.G., and Walker, E.J. 1982. The rate and extent of reaction between calcium hydroxide and pulverized fuel ash. In *Proceedings of the Use of PFA in Concrete*, Leeds, April, Vol. 1, pp. 71–81.

Caldarone, M.A., Taylor, P.C., Detwiler, R.J., and Bhidé, S.B. 2005. *Guide specification for high-performance concrete for bridges*. EB233, 1st ed. Portland Cement Association, Skokie, IL.

Carette, G., Bilodeau, A., Chevrier, R.L., and Malhotra, V.M. 1993. Mechanical properties of concrete incorporating high volumes of fly ash from sources in the U.S. *ACI Materials Journal*, 90(6), 535–544.

Carette, G.G., and Malhotra, V.M. 1983. Mechanical properties, durability, and drying shrinkage of portland cement concrete incorporating silica fume. *Cement, Concrete, and Aggregates*, 5(1), 3–13.

Cochran, J.W., and Boyd, T.J. 1993. Beneficiation of fly ash by carbon burnout. In *Proceedings of the Tenth International Ash Use Symposium*, EPRI TR 101774, Vol. 2, Paper 73, pp. 73-1 to 73-9.

Concrete Society. 1991. *The use of GGBS and PFA in concrete*. Technical Report 40. Concrete Society, Wexham, Slough.

CSA A3001. 2003. "Cementitious Materials Compondium." CSA A3001, Canadian Standards Association, Mississauga, Ontario, Canada.

Davies, R.E. 1954. Pozzolanic materials—With special reference to their use in concrete pipe. Technical Memo. American Concrete Pipe Association. Irving, TX.

Davis, R.E., Carlson, R.W., Kelly, J.W., and Davis, H.E. 1937. Properties of cements and concretes containing fly ash. *Journal of the American Concrete Institute*, 33, 577–611.

Demoulian, E., Gourdin, P., Hawrhorn, F., and Vernet, C. 1980. Influence of slags chemical composition and texture on their hydraulicity (in French). In *Proceedings of the Seventh International Congress on the Chemistry of Cement*, Paris, Vol. 2, pp. 89–94.

Diamond, S. 1983a. On the glass present in low-calcium and high-calcium flyashes. *Cement and Concrete Research*, 13(4), 459–464.

Diamond, S. 1983b. Alkali reactions in concrete—Pore solution effects. In *Proceedings of 6th International Conference on Alkalis in Concrete* (ed. G.M. Idorn and S. Rostam). Danish Concrete Association, Copenhagen, pp. 155–166.

Diamond, S. 2000. Mercury porosimetry: An inappropriate method for the measurement of pore size distributions in cement-based materials. *Cement and Concrete Research*, 30, 1517–1525.

Diamond, S., and Penko, M. 1992. Alkali silica reaction processes: The conversion of cement alkalis to alkali hydroxide. In *Durability of Concrete—G.M. Idorn International Symposium* (ed. Jens Holm). ACI SP-131. American Concrete Institute, Detroit, pp. 153–168.

Dunstan, E.R. 1980. A possible method for identifying fly ashes that will improve the sulfate resistance of concretes. *Cement, Concrete and Aggregates*, pp. 20–30.

Dunstan, E.R. 1984. *Fly ash and fly ash concrete*. Report REC-ERC-82-1. U.S. Bureau of Reclamation, Denver.

Efstathiadis, E. 1978. Greek concrete of three millenniums. Technical Report. Research Center of the Hellenic Ministry of Public Works, Athens.

Elfert, R.J. 1974. *Bureau of reclamation experiences with flyash and other pozzolans in concrete*. Information Circular 8640. U.S. Bureau of Mines, Washington, DC, pp. 80–93.

Ellis-Don Construction, Ltd. 1996. Leading edge. *Edifice Magazine*, March, p. 14.

Fidjestol, P., and Lewis, R. 1998. Microsilica as an addition. In *Lea's chemistry of cement and concrete* (ed. P.C. Hewlett). Arnold, London, pp. 675–708.

Fiskaa, O.M. 1973. *Betong i Alunskifer*. Publication 101. Norsk Geoteknisk Institutt, Oslo, Norway.

FitzGibbon, M.E. 1976a. Large pours for reinforced concrete structures. *Concrete*, 10(3), 41.

FitzGibbon, M.E. 1976b. Large pours—2, heat generation and control. *Concrete*, 10(12), 33–35.

Folliard, K.J., and Sandberg, P. 1994. Mechanisms of concrete deterioration by sodium sulfate crystallization. ACI Special Publication (SP-145). In *Third International ACI/CANMET Conference on Concrete Durability*, Nice, France, pp. 933–946.

Fournier, B., Nkinamubanzi, P.-C., and Chevrier, R. 2004. Comparative field and laboratory investigations on the use of supplementary cementing materials to control alkali-silica reaction in concrete. In *Proceedings of the Twelfth International Conference Alkali-Aggregate Reaction in Concrete* (ed. T. Mingshu and D. Min). Vol. 1. International Academic Publishers/World Publishing Corp., Beijing, pp. 528–537.

Gebler, S., and Klieger, P. 1983. Effect of fly ash on the air-void stability of concrete. In *Proceedings of the 1st International Conference on the Use of Fly Ash, Silica Fume, Slag and Other Mineral By-Products in Concrete* (ed. V.M. Malhotra). ACI SP-79, Vol. 1. American Concrete Institute, Farmington Hills, MI, pp. 103–142.

Gebler, S.H., and Klieger, P. 1986. *Effect of fly ash on some of the physical properties of concrete.* Research and Development Bulletin RD089. Portland Cement Association, Skokie, IL.

Ghorab, H.Y., Heinz, D., Ludwig, U., Meskendahl, T., and Wolter, A. 1980. On the stability of calcium aluminate sulphate hydrates in pure systems and in cements. In *Proceedings of the 7th International Congress on the Chemistry of Cement,* Paris, Vol. 4, pp. 496–503.

Ghosh, R.S., and Timusk, J. 1981. Creep of fly ash concrete. *ACI Materials Journal,* 78(5), 351–357.

Gilliland, J.L., and Moran, W.T. 1949. Siliceous admixture for the Davis Dam. *Engineering News Record,* February 3, p. 62.

Groppo, J. 2001. The recovery of high quality fuel from ponded utility coal combustion ash. In *ACCA Fourteenth International Symposium on Coal Combustion Products,* San Antonio, TX, Vol. 1, p. 14-1.

Gruber, K.A., Ramlochan, T., Boddy, A., Hooton, R.D., and Thomas, M.D.A. 2001. Increasing concrete durability with high-reactivity metakaolin. *Cement and Concrete Composites,* 23(6), 479–484.

Hassan, K.E., and Cabrera, J.G. 1998. The use of classified fly ash to produce high performance concrete. In *Sixth CANMET/ACI International Conference on Fly Ash, Silica Fume, Slag, and Natural Pozzolans in Concrete* (ed. V.M. Malhotra). SP-178. American Concrete Institute, Farmington Hills, MI, pp. 21–36.

Haynes, H., O'Neill, R., and Mehta, P.K. 1996. Concrete deterioration from physical attack by salts. *Concrete International,* 18(1), 63–68.

Helmuth, R. 1987. *Fly ash in cement and concrete.* Portland Cement Association, Skokie, IL.

Hinrichs, W., and Odler, I. 1989. Investigations on the hydration of portland blast-furnace slag cement. *Advances in Cement Research,* 2, 9–13, 15–20.

Hobbs, D.W. 1986. Carbonation of concrete containing PFA. *Magazine of Concrete Research,* 40(143).

Holland, T.C., Krysa, A., Luther, M., and Liu, T. 1986. Use of silica-fume concrete to repair abrasion-erosion damage in the Kinzua Dam stilling basin. In *Proceedings of the Second CANMET/ACI International Conference on Fly Ash, Silica Fume, Slag, and Natural Pozzolans in Concrete* (ed. V.M. Malhotra). SP-91, Vol. 2. American Concrete Institute, Farmington Hills, MI, pp. 841–864.

Hooton, R.D. 1987. The reactivity and hydration products of blast-furnace slag. In *Supplementary cementing materials for concrete* (ed. V.M. Malhotra). CANMET, Ottawa, pp. 247–288.

Hooton, R.D. 1993. Influence of silica fume replacement of cement on physical properties and resistance to sulfate attack, freezing and thawing, and alkali-silica reactivity. *ACI Materials Journal,* 90(2), 143–151.

Hooton, R.D., and Emery, J.J. 1990. Sulfate resistance of a Canadian slag. *ACI Materials Journal,* 87(6), 547–555.

Hooton, R.D., Thomas, M.D.A., and Ramlochan, T. 2008. Durability of ternary blend concrete with silica fume and blastfurnace slag: Laboratory and outdoor exposure site studies. In *Proceedings of 13th International Conference on Alkali-Aggregate Reaction,* Trondheim, Norway, CD-ROM.

Idorn, G.M. 1997. *Concrete progress from antiquity to the third millennia.* Thomas Telford, London.

Irassar, E.F., Di Maio, A., and Batic, O.R. 1996. Sulfate attack on concrete with mineral admixtures. *Cement and Concrete Research*, 26(1), 113–123.

Jiang, Z., Sun, Z., and Wang, P. 2005. Autogenous relative humidity change and autogenous shrinkage of high-performance cement pastes. *Cement and Concrete Research*, 35, 1539–1545.

Klieger, P., and Isberner, A.W. 1967. Laboratory studies of blended cement—Portland blast-furnace slag cements. *Journal, PCA Research and Development Department Laboratories*, 9(3), 2–22.

Kokubu, M.M. 1969. Fly ash and fly ash cement. In *Proceedings of 5th International Symposium on the Chemistry of Cement*. Cement Association of Japan, Part IV, pp. 75–105, Tokyo, Japan.

Kosmatka, S.H., and Wilson, M.L. 2011. *Design and control of concrete mixtures*. EB001, 15th ed. Portland Cement Association, Skokie, IL.

Kostuch, J.A., Walters, G.V., and Jones, T.R. 1993. High performance concretes incorporating metakaolin—A review. In *Proceedings of Concrete 2000* (ed. R.K. Dhir and M.R. Jones). Vol. 2. University of Dundee, pp. 1799–1811, Dundee, Scotland.

Lane, D.S., and Ozyildirim, C. 1999. *Combinations of pozzolans and ground, granulated, blast-furnace slag for durable hydraulic cement concrete*. VTRC 00-R1. Virginia Transportation Research Council, Charlottesville, VA.

Langley, W.S., Carette, G.G., and Malhotra, V.M. 1992. Strength development and temperature rise in large concrete blocks containing high volumes of low-calcium (ASTM class F) fly ash. *ACI Materials Journal*, 89(4), 362–368.

Langley, W.S., and Leaman, G.H. 1998. Practical uses of high-volume fly ash concrete utilizing a low calcium fly ash. In *Proceedings of the Sixth CANMET/ACI/JCI International Conference on Fly Ash, Silica Fume, Slag and Natural Pozzolans in Concrete* (ed. V.M. Malhotra). ACI SP-178, Vol. 1. American Concrete Institute, Detroit, pp. 545–574.

Lea, F.M. 1971. *The chemistry of cement and concrete*. Chemical Publishing Co., New York.

Lee, K.M., Lee H.K., Lee, S.H., and Kim, G.Y. 2006. Autogenous shrinkage of concrete containing granulated blast-furnace slag. *Cement and Concrete Research*, 36(7), 1279–1285.

Lewis, R., Sear, L., Wainwright, P., and Ryle, R. 2003. Cementitious additions. In *Advanced concrete technology: Constituent materials* (ed. J. Newman and B.S. Choo). Elsevier, Oxford, pp. 3/1–3/66.

Li, H., Wee, T.H., and Wong, S.F. 2002. Early-age creep and shrinkage of blended cement concrete. *ACI Materials Journal*, 99(1), 3–10.

Locher, F.W. 2006. *Cement: Principles of production and use*. Verlag Bau+Technik GmbH, Düsseldorf.

Longuet, P., Burglen, L., and Zelwer, A. 1973. La phase liquide du cement hydrate. *Revue Materiaux de Construction et travaux Publics*, 676, 35–41.

Luther, M., and Hansen, W. 1989. Comparison of creep and shrinkage of high-strength silica fume concretes with fly ash concretes of similar strength. In *Proceedings, of CANMET/ACI Third International Conference on the Use of Fly Ash, Silica Fume, Slag, and Natural Pozzolans in Concrete* (ed. V.M. Malhotra). SP-114, Vol. 1. American Concrete Institute, Detroit, pp. 573–591.

Malhotra, V.M., and Mehta, P.K. 1996. *Pozzolanic and cementitious materials, advances in concrete technology*. Vol. 1. Gordon and Breach Publishers, Amsterdam.

Malhotra, V.M., and Mehta, P.K. 2005. *High-performance, high-volume fly ash concrete*. Supplementary Cementing Materials for Sustainable Development, Ottawa, Canada.

Malhotra, V.M., Ramachandran, V.S., Feldman, R.F., and Aïtcin, P.C. 1987. *Condensed silica fume in concrete*. CRC Press, Boca Raton, FL.

Malinowski, R., and Frifelt, K. (in cooperation with Bonits, Hjerthem, and Flodin). 1993. *Prehistoric hydraulic mortar: The Ubaid period 5–4000 years BC: Technical properties*. Document D12. Swedish Council for Building Research, Stockholm, Sweden.

Manmohan, D., and Mehta, P.K. 1981. Influence of pozzolanic, slag, and chemical admixtures on pore size distribution and permeability of hardened cement pastes. *Cement, Concrete and Aggregates*, 3(1), 63–67.

Massazza, F. 1998. Pozzolana and pozzolanic cements. In *Lea's chemistry of cement and concrete* (ed. P.C. Hewlett). Arnold, London, pp. 471–631.

Mather, B. 1957. Laboratory tests of portland blast-furnace slag cements. *ACI Journal, Proceedings*, 54(3), 205–232.

Matthews, J.D. 1984. Carbonation of 10-year-old concretes with and without pulverized-fuel ash. In *Ashtech '84*, London, p. 398A.

McCarthy, G.J., Solem, J.K., Manz, O.E., and Hassett, D.J. 1990. Use of a database of chemical, mineralogical and physical properties of North American fly ash to study the nature of fly ash and its utilization as a mineral admixture in concrete. In *MRS Symposium on Fly Ash and Coal Conversion By-Products Characterization, Utilization and Disposal V* (ed. R.L. Day and F.P. Glasses). Vol. 178. Materials Research Society, Pittsburgh, PA, pp. 3–33.

Mehta, P.K. 1981. Studies on blended portland cements containing santorin-earth. *Cement and Concrete Research*, 11, 507–518.

Mehta, P.K. 1983. Pozzolanic and cementitious byproducts as mineral admixtures for concrete—A critical review. In *Proceedings of the First International Conference on the Use of Fly Ash, Silica Fume, Slag and Other Mineral By-Products in Concrete*. ACI SP-79, Vol. 1. America Concrete Institute, Detroit, pp. 1–46.

Mehta, P.K. 1986. Effect of fly ash composition on sulfate resistance of cement. *ACI Materials Journal*, November–December, pp. 994–1000.

Mehta, P.K. 1987. Natural pozzolans. In *Supplementary cementing materials for concrete* (ed. V.M. Malhotra). CANMET-SP-86-8E. Canadian Government Publishing Center, Supply and Services, Ottawa, Canada, pp. 1–33.

Mehta, P.K. 1992. Rice husk ash—A unique supplementary cementing material. In *CANMET, Proceedings of the International Symposium on Advances in Concrete Technology*, Athens, Greece (ed. V.M. Malhotra), May, pp. 407–430.

Mehta, P.K. 2002. Performance of high-volume fly ash concrete in hot weather. In *Innovations in design with emphasis on seismic, wind, and environmental loading: Quality control and innovations in materials/hot-weather concreting* (ed. V.M. Malhotra). ACI SP-209. American Concrete Institute, Farmington Hills, MI, pp. 47–52.

Mehta, P.K., and Langley, W.S. 2000. Monolith foundation, built to last a 1000 years. *Concrete International*, July, pp. 27–32.

Meininger, R.C. 1981. *Use of fly ash in air-entrained concrete. Report of recent NSGA-NRMCA research laboratory studies*. National Ready Mixed Concrete Association, Silver Spring, MD.

Mindess, S., Young, J.F., and Darwin, D. 2003. *Concrete*. 2nd ed. Prentice Hall, Englewood Cliffs, NJ.

Monk, M. 1973. Portland PFA cement: A comparison between intergrinding and blending. *Magazine of Concrete Research*, 35(124), 131–141.

Moranville-Regourd, M. 1998. Cements made from blastfurnace slag. In *Lea's chemistry of cement and concrete* (ed. P.C. Hewlett). Arnold, London, pp. 633–674.

Mustard, J.N., and MacInnis, C. 1959. The use of fly ash in concrete by Ontario Hydro. *Engineering Journal*, December, pp. 74–79.

Naik, T.R., Ramme, B.W., and Tews, J.H. 1995. Pavement construction with high-volume class C and class F fly ash concrete. *ACI Materials Journal*, 92(2), 200–210.

Nixon, P.J., and Page, C.L. 1987. Pore solution chemistry and alkali aggregate reaction. In *Concrete Durability, Katherine and Bryant Mather International Conference* (ed. J.M. Scanlon). ACI SP-100, Vol. 2. American Concrete Institute, Detroit, pp. 1833–1862.

Obla, K.H., Hill, R.L., Thomas, M.D.A., Shashiprakash, S.G., and Perebatova, O. 2003. Properties of concrete containing ultra-fine fly ash. *ACI Materials Journal*, 100, 426–433.

Odler, I. 2000. *Special inorganic cements*. E&FN Spon, London.

Ogawa, K., Uchikawa, H., Takemoto, K., and Yasui, I. 1980. The mechanism of the hydration in the system C_3S pozzolanas. *Cement and Concrete Research*, 10(5), 683–696.

Osborne, G.J. 1986. Carbonation of blastfurnace slag cement concrete. *Durability of Building Materials*, 4, 81–96.

Osborne, G.J. 1989. Carbonation and permeability of blast furnace slag cement concretes from field structures. In *Proceedings of 3rd CANMET/ACI International Conference on Fly Ash, Silica Fume, Slag and Natural Pozzolans in Concrete* (ed. V.M. Malhotra). ACI SP-114, Vol. 2. American Concrete Institute, Detroit, pp. 1209–1237.

Osborne, G.J., and Connell, M.D. 2000. Effects of cementitious slag content and aggregate type on properties and durability of portland blast furnace slag cement concretes. In *Proceedings of the 5th CANMET/ACI International Conference on the Durability of Concrete*. ACI SP-192, Vol. 1. American Concrete Institute, Detroit, pp. 119–139.

Owens, P.L. 1979. Fly ash and its usage in concrete. *Concrete Magazine*, July, pp. 22–26.

Parrott, L.J. 1987. *A review of carbonation in reinforced concrete*. Building Research Establishment, Garston, UK.

Pigeon, M., and Pleau, R. 1995. *Durability of concrete in cold climates*. E&FN Spon, London.

Pistilli, M.F. 1983. Air-void parameters developed by air-entraining admixtures as influenced by soluble alkalis from fly ash and portland cement. *ACI Journal, Proceedings*, 80(3), 217–222.

Portland Cement Association (PCA). 2002. *Types and causes of concrete deterioration*. PCA Report IS536. Portland Cement Association, Skokie, IL.

Powers, T.C. 1945. A working hypothesis for further studies of frost resistance of concrete. *ACI Journal, Proceedings*, 41(3), 245–272.

Powers, T.C. 1961. *Physical properties of cement paste*. Bulletin 154. Research Laboratory of the Portland Cement Association, Skokie, IL.

Powers, T.C., and Brownyard, T.L. 1948. *Studies of the physical properties of hardened portland cement paste*. Bulletin 22. Research Laboratory of the Portland Cement Association, Skokie, IL.

Price, W.H. 1975. Pozzolans—A review. *ACI Journal, Proceedings*, 72(5), 225–232.

Prusinski, J.R. 2006. Slag as a cementitious material. In *Significance of tests and properties of concrete and concrete-making materials* (ed. J.F. Lamond and J.H. Pielert). ASTM STP 169D. American Society of Testing and Materials, West Conshohocken, PA, pp. 512–530.

Radjy, F.F., Bogen, T., Sellevold, E.J., and Loeland, K.E. 1986. A review of experiences with condensed silica-fume concretes and products. In *Proceedings of the 2nd International Conference on Fly Ash, Silica Fume, Slag and Natural Pozzolans in Concrete* (ed. V.M. Malhotra). ACI SP-91, Vol. 2. American Concrete Institute, Detroit, pp. 1135–1152.

Ramlochan, T., and Thomas, M.D.A. 2000. The effect of metakaolin on external sulphate attack. In *Proceedings of the 4th ACI/CANMET International Conference on the Durability of Concrete* (ed. V.M. Malhotra), ACI SP-192, Vol. 1, pp. 239–251.

Ramlochan, T., Thomas, M.D.A., and Gruber, K.A. 2000. The effect of metakaolin on alkali-silica reaction in concrete. *Cement and Concrete Research*, 30, 339–344.

Ramlochan, T., Thomas, M.D.A., and Hooton, R.D. 2004. The effect of pozzolans and slag on the expansion of mortars cured at elevated temperature. Part II. Microstructural and microchemical investigations. *Cement and Concrete Research*, 34, 1341–1356.

Ramlochan, T., Zacarias, P., Thomas, M.D.A., and Hooton, R.D. 2003. The effect of pozzolans and slag on expansion of mortars cured at elevated temperature. Part I. Expansive behaviour. *Cement and Concrete Research*, 33, 807–814.

Riding, K.A., Poole, J.L., Schindler, A.K., Juenger, M.C.G., and Folliard, K.J. 2008. Quantification of effects of fly ash type on concrete early-age cracking. *ACI Materials Journal*, 105(2), 149–155.

Rivest, M., Bouzoubaa, N., and Malhotra, V.M. 2004. Strength development and temperature rise in high-volume fly ash and slag concretes in large experimental monoliths. In *Proceedings of the 8th CANMET/ACI International Conference on Fly Ash, Silica Fume, Slag, and Natural Pozzolans in Concrete* (ed. V.M. Malhotra). ACI SP-221. American Concrete Institute, Detroit, pp. 859–878.

Roberts, L.R., and Taylor, P.C. 2007. Understanding cement-SCM-admixture interaction issues: Staying out of the safety zone. *Concrete International*, January, pp. 33–41.

Roy, D.M., and Parker, K.M. 1983. Microstructures and properties of granulated slag-portland cement blends at normal and elevated temperatures. In *Proceedings of the 1st International Conference on the Use of Fly Ash, Silica Fume, Slag and Other Mineral By-Products in Concrete* (ed. V.M. Malhotra). SP-79, Vol. 1. American Concrete Institute, Farmington Hills, MI, pp. 397–414.

Ryell, J., and Bickley, J.A. 1987. Scotia Plaza: High strength concrete for tall buildings. In *Proceedings of the Symposium on Utilization of High Strength Concrete*, Stavanger, Norway, pp. 641–655.

Saad, M.N.A., de Andrade, W.P., and Paulon, V.A. 1982. Properties of mass concrete containing an active pozzolan made from clay. *Concrete International*, July, pp. 59–65.

Sabanegh, N., Gao, Y., Suuberg, E., and Hurt, R. 1997. Interaction of coal fly ash with concrete surfactants: Diffusional transport and adsorption. In *Proceedings of the 9th International Conference on Coal Science*. Vol. 3. P&W Druck and Verlag, Essen, Germany, pp. 1907–1910.

Schlorholtz, S. 2006. Supplementary cementitious materials. In *Significance of tests and properties of concrete and concrete-making materials* (ed. J.F. Lamond and J.H. Pielert). ASTM STP 169D. American Society of Testing and Materials, West Conshohocken, PA, pp. 495–511.

Sear, L.K.A. 2001. *The properties and use of coal fly ash*. Thomas Telford, London.

Sellevold, E.J., and Nilsen, T. 1987. Condensed silica fume in concrete: A world review. In *Supplementary cementing materials for concrete* (ed. V.M. Malhotra). CANMET, Ottawa, pp. 165–243.

Sellevold, E.J., and Radjy, F.F. 1983. Condensed silica fume (microsilica) in concrete: Water demand and strength development. In *Proceedings of the 1st International Conference on the Use of Fly Ash, Silica Fume, Slag and Other Mineral By-Products in Concrete* (ed. V.M. Malhotra). ACI SP-79, Vol. 2. American Concrete Institute, Detroit, pp. 677–694.

Shashiprakash, S.G., and Thomas, M.D.A. 2001. Sulfate resistance of mortars containing high-calcium fly ashes and combinations of highly reactive pozzolans and fly ash. In *Proceedings of the Seventh CANMET/ACI International Conference on Fly Ash, Silica Fume, Slag and Natural Pozzolans in Concrete* (ed. VM Malhotra). ACI SP-199. American Concrete Institute, Detroit.

Shehata, M.H., and Thomas, M.D.A. 2000. The effect of fly ash composition on the expansion of concrete due to alkali silica reaction. *Cement and Concrete Research*, 30(7), 1063–1072.

Shehata, M.H., and Thomas, M.D.A. 2002. Use of ternary blends containing silica fume and fly ash to suppress expansion due to alkali-silica reaction in concrete. *Cement and Concrete Research*, 32(3), 341–349.

Shehata, M.H., Thomas, M.D.A., and Bleszynski, R.F. 1999. The effects of fly ash composition on the chemistry of pore solution in hydrated cement pastes. *Cement and Concrete Research*, 29, 1915–1920.

Skalny, J., Marchand, L., and Odler, I. 2002. *Sulfate attack on concrete*. Spon Press, London.

Skrastins, J.I., and Zoldners, N.G. 1983. Ready-mixed concrete incorporating condensed silica fume. In *Proceedings of the 1st International Conference on the Use of Fly Ash, Silica Fume, Slag and Other Mineral By-Products in Concrete* (ed. V.M. Malhotra). ACI SP-79, Vol. 2. American Concrete Institute, Detroit, pp. 813–829.

Slag Cement Association (SCA). 2006. www.slagcement.org.

Smith, I.A. 1967. The design of fly ash concretes. In *Proceedings of the Institute of Civil Engineers*, London, Vol. 36, pp. 769–790.

Smolczyk, H.G. 1980. Slag structure and identification of slags. In *Proceedings of the Seventh International Congress on the Chemistry of Cement*, Paris, Vol. 1, No. III, pp. 3–17.

Stanish, K.D., Hooton, R.D., and Thomas, M.D.A. 2001. *Prediction of chloride penetration in concrete*. Publication FHWA-RD-00-142. U.S. Department of Transportation, Federal Highway Administration, Washington, DC.

Stanton, T.E. 1940. Expansion of concrete through reaction between cement and aggregate. *Proceedings of the American Society of Civil Engineers*, 66(10), 1781–1811.

Stanton, T.E. 1952. Studies of use of pozzolans for counteracting excessive concrete expansion resulting from reaction between aggregates and the alkalies in cement. In *Pozzolanic materials in mortars and concretes*. ASTM STP 99. American Society for Testing and Materials, Philadelphia, pp. 178–203.

Stark, D. 2002. *Performance of concrete in sulfate environments*. PCA R&D Bulletin RD129. Portland Cement Association, Skokie, IL.

Sturrup, V.R., Hooton, R.D., and Clendenning, T.G. 1983. Durability of fly ash concrete. In *Proceedings of the 1st International Conference on the Use of Fly Ash, Silica Fume, Slag and Other Mineral By-Products in Concrete* (ed. V.M. Malhotra). ACI SP-79, Vol. 1. American Concrete Institute, Farmington Hills, MI, pp. 71–86.

Taylor, H.F.W. 1997. *Cement chemistry*. 2nd ed. Thomas Telford, London.

Tazawa, E., and Miyazawa, S. 1995. Experimental study on mechanism of autogenous shrinkage of concrete. *Cement and Concrete Research*, 25(8), 1633–1638.

Tennis, P.D., and Jennings, H.M. 2000. A model for two types of calcium silicate hydrate in the microstructure of portland cement pastes. *Cement and Concrete Research*, Piscataway, NJ, 3, 855–863.

Thomas, A. 1979. Metallurgical and slag cements, the indispensable energy savers. Presented at General Practices, IEEE Cement Industry 21 Technical Conference.

Thomas, M.D.A. 1989. The effect of curing on the hydration and pore structure of hardened cement paste containing pulverized-fuel ash. *Advances in Cement Research*, 2(8), 181–188.

Thomas, M.D.A. 1997. Laboratory and field studies of salt scaling in fly ash concrete. In *Frost resistance of concrete* (ed. M.J. Setzer and R. Auberg). EeFN spon, Essen, Germany.

Thomas, M.D.A. 2007. *Optimizing the use of fly ash in concrete*. PCA IS548. Portland Cement Association, Skokie, IL.

Thomas, M.D.A. 2011. The effect of supplementary cementing materials on alkali-silica reaction: A review. *Cement and Concrete Research*, 41, 1224–1231.

Thomas, M.D.A., and Bamforth, P.B. 1999. Modelling chloride diffusion in concrete: Effect of fly ash and slag. *Cement and Concrete Research*, 29, 487–495.

Thomas, M.D.A., and Bleszynski, R.F. 2001. The use of silica fume to control expansion due to alkali-aggregate reactivity in concrete—A review. In *Materials science of concrete VI* (ed. S. Mindess and J. Skalny). American Ceramics Society, Westerville, OH, pp. 377–434.

Thomas, M.D.A., Cail, K., and Hooton, R.D. 1998. Development and field applications of silica fume concrete in Canada—A retrospective. *Canadian Journal of Civil Engineering*, 25, 391–400.

Thomas, M.D.A., Folliard, K., Drimalas, T., and Ramlochan, T. 2008. Diagnosing delayed ettringite formation in concrete structures. *Cement and Concrete Research*, 38, 841–847.

Thomas, M.D.A., Hopkins, D.S., Perreault, M., and Cail, K. 2007. Ternary cement in Canada. *Concrete International*, 29(7), 59–64.

Thomas, M.D.A., and Innis, F.A. 1998. Effect of slag on expansion due to alkali-aggregate reaction in concrete. *ACI Materials Journal*, 95(6).

Thomas, M.D.A., and Matthews, J.D. 1994. *Durability of PFA concrete*. BRE Report BR216. Building Research Establishment (Department of the Environment), Watford, UK.

Thomas, M.D.A., Matthews, J.D., and Haynes, C.A. 2000. Carbonation of fly ash concrete. In *Proceedings of the 4th ACI/CANMET International Conference on the Durability of Concrete* (ed. V.M. Malhotra). ACI SP-192, Vol. 1, pp. 539–556.

Thomas, M.D.A., Mukherjee, P.K., Sato, J.A., and Everitt, M.F. 1995. Effect of fly ash composition on thermal cracking in concrete. In *Proceedings of the Fifth CANMET/ACI International Conference on Fly Ash, Silica Fume, Slag, and Natural Pozzolans in Concrete* (ed. V.M. Malhotra). SP-153, Vol. 1. American Concrete Institute, Farmington Hills, MI, pp. 81–98.

Thomas, M.D.A., and Shehata, M. 2004. Use of blended cements to control expansion of concrete due to alkali-silica reaction. In *Supplementary papers of the 8th CANMET/ACI International Conference of Fly Ash, Silica Fume, Slag and Natural Pozzolans in Concrete*, Las Vegas, NV, pp. 591–607.

Thomas, M.D.A., Shehata, M., and Shashiprakash, S.G. 1999. The use of fly ash in concrete: Classification by composition. *Cement, Concrete and Aggregates*, 12(2), 105–110.

Thomas, M.D.A., and Skalny, J. 2006. Chemical resistance of concrete. In *Significance of tests and properties of concrete and concrete making materials* (ed. J.F. Lamond and J.H. Pielert). ASTM STP 169D. American Society for Testing and Materials, West Conshohocken, PA, pp. 253–273.

Tikalsky, P.J., and Carrasquillo, R.L. 1992. Influence of fly ash on the sulfate resistance of concrete. *ACI Materials Journal*, January–February, pp. 69–75.

Traetteberg, A. 1978. Silica fume as pozzolanic materials. *Il Cemento*, 75, 369–375.

Tuutti, K. 1982. *Corrosion of steel in concrete*. Report 4-82. Swedish Cement and Concrete Research Institute, Borås, Sweden.

Uchikawa, H. 1986. Effect of blending component on hydration and structure formation. In *Proceedings of the 8th International Congress on the Chemistry of Cement*, Rio de Janeiro, Vol. 1, pp. 249–280.

U.S. Bureau of Reclamation (USBR). 1948. *Physical and chemical properties of fly ash—Hungry horse dam*. Laboratory Report CH-95. U.S. Bureau of Reclamation, Washington, DC.

von Fay, K.F., and Pierce, J.S. 1989. Sulfate resistance of concretes with various fly ashes. *ASTM Standardization News*, December, pp. 32–37.

Wainwright, P.J. 1986. Properties of fresh and hardened concrete incorporating slag cements. In *Cement replacement materials* (ed. R.N. Swamy). Concrete Technology Design Vol. 3. Surrey University Press, Guildford, Surrey, UK, pp. 100–133.

Wainwright, P.J., and Tolloczko, J.J.A. 1986. Early and later age properties of temperature cycled slag-OPC concrete. In *Proceedings of the Second CANMET/ ACI International Conference on Fly Ash, Silica Fume, Slag, and Natural Pozzolans in Concrete* (ed. V.M. Malhotra). ACI SP-91, Vol. 2. American Concrete Institute, Farmington Hills, MI, pp. 1293–1322.

Wang, H., Qi, C., Farzam, H., and Turici, J. 2006. Interaction of materials used in concrete: Effects of fly ash and chemical admixtures on portland cement performance. *Concrete International*, April, pp. 47–52.

Whiting, D.A. 1981. *Rapid determination of the chloride permeability of concrete*. Report FHWA/RD-81/119. Federal Highway Administration, Washington, DC.

Whiting, D. 1989. *Strength and durability of residential concretes containing fly ash*. PCA R&D Bulletin RD099. Portland Cement Association, Skokie, IL.

Williams, J.T., and Owens, P.L. 1982. The implications of a selected grade of United Kingdom pulverized fuel ash on the engineering design and use in structural concrete. In *Proceedings of the International Symposium on the Use of PFA in Concrete* (ed. J.G. Cabrera and A.R. Cusens). Department of Civil Engineering, University of Leeds, Leeds, UK, pp. 301–313.

Young, J.F., Mindess, S., Gray, R.J., and Bentur, A. 1998. *The science and technology of civil engineering materials*. Prentice Hall, Upper Saddle River, NJ.

Zhang, M.H., Tam, C.T., and Leow, M.P. 2003. Effect of water-to-cementitious materials ratio and silica fume on the autogenous shrinkage of concrete. *Cement and Concrete Research*, 33(10), 1687–1694.

Index

Printed and bound by CPI Group (UK) Ltd, Croydon, CR0 4YY

18/10/2024

01776271-0003